THE HUNTERS

THE HUNTERS

Dr Philip Whitfield
Illustrated by Richard Orr

Consultant Editor, John Farrand, Jr
The American Museum of Natural History

Introduction by Desmond Morris

Simon and Schuster
New York

CONTENTS

Text by Dr Philip Whitfield
Major illustrations by Richard Orr
Additional drawings by Michael Woods

Editor: Jinny Johnson
Designer: Julia Clappison
Design assistant: Roger Walton
Editorial assistant: Janet Wilson

The Hunters was edited and designed by
Marshall Editions Limited,
71 Eccleston Square, London SW1V 1PJ
© 1978 Marshall Editions Limited

Published by Simon and Schuster
A Division of Gulf & Western Corporation
Simon & Schuster Building,
Rockefeller Center,
1230 Avenue of the Americas,
New York, New York 10020

Printed and bound in Spain
by Printer industria gráfica sa
Sant Vicenç dels Horts   Barcelona
D.L.B. 24937-1978

Library of Congress Cataloging in
Publication Data

Whitfield, Philip.
  The hunters.

  Bibliography: p.
  Includes index.
  1. Predation (Biology) I. Orr, Richard.
II. Farrand, John. III. Title.
QL758.W45   591.5'3   78–7960
ISBN 0–671–24398–5

INTRODUCTION

by Desmond Morris
Research Fellow, Wolfson College, Oxford

When our remote ancestors abandoned their simple, sweet, fruit-picking existence in the forest trees and turned to a hunting way of life, they opened the door to a success story unparalleled in the whole of animal evolution. Strangely, this success came because our apeman predecessors were physically ill-equipped for the task ahead. Even when they stood up on their hind legs and ran as fast as they could, they were no match for their carnivore competitors, the lions and leopards, the wild dogs and hyenas. At first they had to make do with small, helpless prey, but they possessed a secret weapon—their large primate brains. What they lacked in savage power they made up for with their quick thinking. And as their thinking became even quicker, they took a vital step—they began to co-operate in achieving the kill. Co-operation meant communication and soon their grunts and yelps developed into words; words carrying information. With this new growth of combined effort, they were able to plan their attacks. Tactics and strategy appeared. Weapons were made and improved, tricks and traps developed, until, at last, they could set off as an efficient hunting party and pit themselves against large prey—animals much bigger than themselves. No species was safe from them now, and no carnivore could better them.

These deadly new hunters—the earliest men—were so successful that they soon began to spread over all the land masses of the earth. As the millennia passed, their populations grew and grew until, about ten thousand years ago, their skills led them over a new threshold, to become stock-keepers and stock-breeders. The hunters became farmers. The hunting phase had lasted more than a million years, and had radically changed, not only man's physique, but also his basic patterns of behaviour. Now, in a sudden rush he was racing ahead into an agricultural epoch and his evolution as an animal species could no longer keep pace with his cultural developments.

Even in modern, industrial times, with all his modern inventions around him, he remained, biologically, a hunter at heart. He no longer needed to kill to survive, but the urge to hunt was strong within him. His weapons were so advanced now that he could kill anything in sight. Real hunting persisted as a wasteful sport, decimating the dwindling populations of his ancient animal prey and his early animal competitors. In its place came symbolic hunting. The marksman hit a bull's eye, but no bull died. Bloodless sports grew out of, and largely replaced, the older blood sports. And in many of his new patterns of living, the hunting system survived in hidden forms. The businessmen setting off to make a "killing" in the city, the team of mountaineers striking out to scale another peak, each in their way was acting out the primeval hunting pattern of our species—with the group organized towards a target, the taking of risks, the planning of manoeuvres, the tactics, the skills, the prolonged pursuit of a difficult goal. These are the qualities of human hunters; but no longer letting blood, instead achieving the kill in some abstract, symbolic form.

Ask any man what makes a boring life and he will tell you that it is monotonous, repetitive work, performed hour-in and hour-out, with no climax, no build-up, no danger, no variety. In other words, what makes for a boring human life is the chew-chew-chew of the hoofed animal, the endlessly munching ungulate. Sadly many men are condemned to such an existence in factories, sheds and warehouses, but even they manage a kind of escape, if only in their day-dreams. In their fantasies—and thank goodness we are capable of fantasies—they live out acts of bravery and daring, with thrills and risks of the kind so badly lacking in their dreary daily routine.

So all of us remain hunters to this day, either symbolic hunters or fantasy hunters, achieving ambitious goals and setting ourselves lofty targets. Little wonder that we are fascinated by the real hunters, belonging to other species, that have managed to survive on the scarred face of the globe that man has altered so radically in a few brief millennia. Tragically, there has been much unnecessary despoiling of the natural environment. We no longer need to prove ourselves against animal prey—we know we can win now—our new weapons have turned every other species into a sitting duck, and it has become meaningless to pit ourselves against such comparatively puny targets. Our goals must become even more lofty than they have been before. But, while we aim ourselves at the stars, we should spare a moment to enjoy the wonders of the natural, animal world that has managed to remain on our man-dominated planet. We can afford, more than ever now, a protective, benign attitude to our beleaguered animal companions. There is so much pleasure to be gained, both aesthetic and scientific, from a close study of their activities. In this book, we have a chance to increase this pleasure, by taking a closer look at the forms and movements of those other hunting species, the ones for whom there is no symbolic kill and for whom the original hunting way of life remains a matter of desperate survival. They are magnificent, fascinating animals, as I am sure you will agree, when you turn these pages.

Desmond Morris

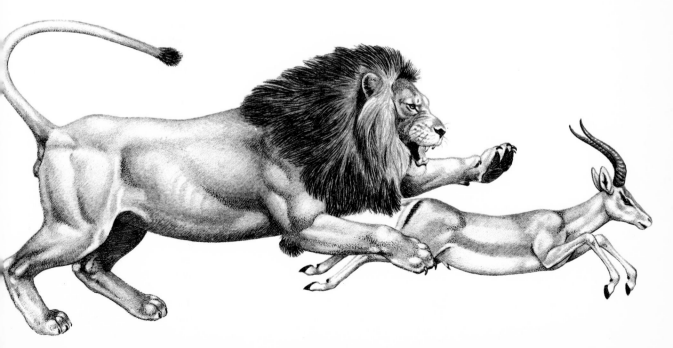

The Hunters: For Ever and Everywhere

For almost as long as life has existed on earth there have been predators. The evolution of the vast variety of today's hunters, killers and eaters of live prey began some 3,000 million years ago—just 1,000 million years after the earth became a solid planet.

The atmosphere of the primitive earth consisted of gases such as nitrogen, carbon dioxide and methane. The environment also contained water, but certainly none of the present-day organisms that demand oxygen for survival could have existed there. Exactly what types of organisms populated this apparently inhospitable world remains a mystery, but the study of micro-fossils and simple creatures of today provides powerful evidence that the earth's first living organisms were probably water-dwelling cells without nuclei, similar in structure to some modern bacteria. These cells could manage without oxygen by feeding on simple organic compounds in the water and generating energy by a process of fermentation.

Even in these early days it is possible that predatory styles of life evolved. Certain bacteria-like cells must have stumbled on the "discovery" that complex organic materials could be obtained by consuming other living organisms. Perhaps these very first hunters were similar to a strange bacterium that exists today but which was only discovered in 1962. The bacterium is called *Bdellovibrio*, and appropriately the first part of its name means leech; for these minute cells, each less than a thousandth of a millimetre long, suck the living contents from other, larger bacteria.

The arrival of blue-green algae

For over 2,000 million years, life on earth consisted entirely of bacteria. The great revolution came with the evolution of a group of plant-like cells, blue-green algae, which transformed the face of the earth. Their magic power was oxygen. Like modern plants the blue-green algae used the pigments in their cells, plus the energy from sunlight, to bind atmospheric carbon dioxide into foodstuffs and incidentally to release oxygen. By this process of photosynthesis the algae turned the earth's atmosphere into the oxygen-rich one of today—but also brought sudden death to most of the fermenting bacteria. For them oxygen was poison.

The challenge of atmospheric oxygen was seemingly the evolutionary "nudge" that stimulated the development of cells with nuclei typical of all higher fungi, plants and animals today. These newly sophisticated cells opened up fantastic possibilities for more complex organisms and from 600 to 700 million years ago the rapid proliferation of new types of organisms has continued unabated.

From single-celled plants and animals to complex mammals, almost every group of living things contains creatures that have become specialized to catch living prey and consume it. Any living plan can, it seems, be modified to produce a machine equipped for a predatory life style.

The lowly fungi, plants that normally digest ready-dead material, can even take on a full predatory life. Most fungi are formed from a cotton wool-like mass of fine threads called hyphae. In the species *Dactylaria gracilis* some of these threads form loops like lassoes. When a small round-worm tries to pass through one of these loops, the ring contracts, trapping the worm. Other hyphae then penetrate the immobilized worm and digest it. Different soil-dwelling fungi shed sticky attacking hyphae, conidia, that adhere to single-celled animals. Each conidium then sends out absorptive hyphae that remove nutrients from the prey.

The single-celled predators

The single-celled aquatic animals known as protozoans, numbering over 30,000 species, reveal a vast array of predatory habits. *Amoeba* species, for example, engulf small prey such as algal cells and bacteria by forming an intucking of their cell membrane into which the food organism is drawn. The intucking eventually forms a closed sac inside the amoeba called a food vacuole. Small bags of digestive enzymes then fuse with the vacuole and the prey is digested and absorbed.

Other protozoans, known as ciliates, have mobile hair-like processes, cilia, on their cell surface which can beat rhythmically back and forth. The currents set up by this beating waft small food organisms into a specialized region of the cell surface where they are taken in and used for food.

Instead of using their cilia for feeding, the predatory protozoans called suctorians develop an array of trapping tentacles. The end of each tentacle is adapted for catching swimming ciliates. As the prey touches on a tentacle tip, specialized attachment and digestive sacs come to the surface of the tentacle and ensnare the prey. Its cell contents are then sucked down the tentacle core and into the predator.

The invertebrate predators

All the many-celled animals of today evolved from the protozoans. Of the groups of animals without backbones—the invertebrates—many, such as the coelenterates, which include the corals, jellyfish and sea anemones, are invariably predatory. Every group member is equipped with elaborate stinging cells which are used to snare and paralyze prey animals in the water. All the sponges, Porifera, feed on minute prey such as bacteria, which they extract by filtering vast volumes of water through a series of ducts and tubes in the body.

Specialist predators abound among the more complex invertebrates. The segmented worms, the group to which the earth-worm belongs, number in their ranks some fearsome burrowing predators which can tunnel rapidly through sand and mud to catch their prey—crustaceans and other worms—with fang-like jaws.

The molluscs, most of which house themselves in protective shells, are a group containing animals that can kill a man. The beautiful cone shell (*Conus*) has an attack proboscis—an elongated part of the mouth—tipped with a single hollow tooth. Into this tooth run toxins that are among the most potent known. A cone shell that dwells on a tropical coral reef uses these paralyzing poisons to kill prey such as fish, but the same toxins injected into a careless swimmer can bring about a rapid death.

Squid and octopuses belong to a group of molluscs often considered to be the most advanced of all invertebrate predators. They have lost their external shells and, using huge eyes, scour the oceans for prey. Creatures such as crabs are actually caught by a deadly embrace from the many, suckered tentacles of these animals.

In terms of sheer numbers, the arthropod animals, which include crustaceans (crabs and the like), insects and arachnids (spiders, scorpions and their relations) comprise most of the multicellular life on earth: there are over 300,000 species of beetle alone. Very many of them are direct predators, but parasites and parasitoids are common too in this group.

Pincer-like front legs are the predatory weapons of the crabs and lobsters. These crustaceans are tough, adaptable predators found in seas throughout the world. Among them is the largest living arthropod predator, the Japanese spider crab, which has an 11 foot (2.3 metre) limb span.

Every one of the 20,000 or more spider species is a predator. Some are far-ranging mobile hunters that overcome their prey by sheer size or the use of poison fangs above the mouth, while others produce and spin silk from secretions in the body into a wide variety of alluring traps and snares.

Insect predators are legion and sometimes both adults and larvae are hunters. Adult dragonflies, for example, hawk for flying insects round river and pond edges and the aquatic larval forms can overpower and eat tadpoles or even small fish with their hinged jaws.

Of the remaining invertebrates, the most familiar are the echinoderms, the group to which starfish, sea urchins and sea cucumbers belong. The starfish are intriguing predators with a novel technique for dealing with well-protected prey such as mussels. The starfish holds the shells of its prey apart using its many suction-pad tube feet then throws its stomach out over the prey like a cloak so that digestion starts immediately outside its body.

The vertebrate predators

Animals with backbones—the vertebrates—are divided into seven distinct classes: the jawless fish (lampreys and hagfish); cartilaginous fish (sharks and rays); bony fish; amphibians; reptiles; birds and mammals. Apart from the lampreys and hagfish, which are either parasites or feed on dead animal remains, all the other classes contain representatives of the specialist hunters that spring rapidly to mind as examples of truly predator beasts.

Such out-and-out carnivores as the large mid-water sharks that snatch fish and other marine creatures in their horrific tooth-studded jaws are cartilaginous fish. So too are the flattened-bodied, bottom-feeding rays that eat molluscs and other organisms from the mud of the sea bed and the enormous 40-foot (12-metre) long basking shark that filters a plankton diet with its gigantic spread of gill rakers. The pike and piranha fish of fresh waters and the marine barracuda are examples of predatory excellence among bony fish. Active swimmers, capable of rapid acceleration in diving to attack prey, they grapple with their victims using sharp teeth.

Newts, salamanders, frogs and toads are the best known amphibians, the most primitive class of land-living vertebrates. All are predators with a predominantly insect diet, and many have sticky tongues which can be flicked out to ensnare passing prey.

The reptiles were the first vertebrates to become totally adapted to a terrestrial life. They have dry, scaly skins which contrast with the moist, unscaled skins of amphibian animals. Reptilian household names include the turtles, tortoises, lizards, snakes and crocodiles. Except for the vegetarian tortoises and seaweed-eating marine iguana of the Galapagos islands, all reptiles are carnivores. The snakes, deeply rooted in man's subconscious as the most evil creatures of all time, are in fact beautifully adapted predators. The bodies of some of these legless reptiles can twine round and crush prey and their lower jaws have flexible ligaments and joints which enable them to swallow prey larger than themselves. Many snakes also have poison-injecting fangs among their predatory armour.

From the reptiles evolved the birds, which first took to the air about 150 million years ago. Talons and hooked beaks are the prime offensive weapons of the owls, hawks, eagles and falcons, which have become adapted to kill and eat mammals and other birds. Many smaller birds are insect eaters. Among them, swifts and nightjars are adapted to feed on the wing, tits and warblers to feed in vegetation, flycatchers and bee-eaters to catch insects during flights from a perch. Fish are caught by diving from the air by gannets and from the water by pelicans. Herons stand in the water to fish and penguins, diving ducks and anhingas plunge under water to find a meal. The long-beaked wading birds have made a speciality of probing in marine mud and sand for soft-bodied invertebrate food.

The bats, flying predators that hunt insects at night using an ultrasonic radar system; the killer whale, seals and dolphins which are streamlined underwater killers, plus a vast range of terrestrial predators share, with man, a place in the class of warm-blooded mammals. Examples of terrestrial hunters include the burrowing moles and insectivorous hedgehogs and shrews, but it is in the carnivora group of mammals that the most photogenic and best studied predators in the world are found—the pack-hunting dogs and wolves; the flesh-eating big cats such as lions, tigers and leopards, and the smaller but equally voracious weasels, minks and martens.

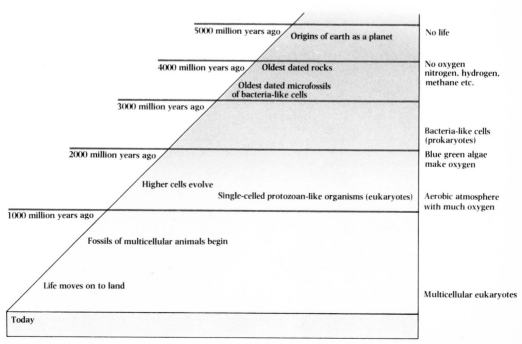

The chronology of the development of life on earth in relation to the history of the earth as a planet, the changes in the earth's atmosphere, and the appearance of different forms of organism, are all detailed on the time chart, right. Vital stages in the sequence of events are the beginning of prokaryotic (bacteria-like) life: and the oxygenation of the earth's atmosphere by prokaryotic blue-green algae. The next stage was the development of eukaroytic life (nucleated cells) which then expanded into increasingly complex multi-cellular organisms—life as it is today.

5000 million years ago — Origins of earth as a planet — No life

4000 million years ago — Oldest dated rocks — No oxygen nitrogen, hydrogen, methane etc.

Oldest dated microfossils of bacteria-like cells

3000 million years ago

Bacteria-like cells (prokaryotes)

2000 million years ago — Blue green algae make oxygen

Higher cells evolve

Single-celled protozoan-like organisms (eukaryotes) — Aerobic atmosphere with much oxygen

1000 million years ago

Fossils of multicellular animals begin

Life moves on to land — Multicellular eukaryotes

Today

The Armoury of the Killers

Two areas of specialization are prominent among predators: sensory system adaptations that enable the hunter to find and identify the prey, and specific designs of offensive weaponry for the kill itself. Both of these demonstrate the extraordinary adaptability of living creatures to undertake unlikely lifestyles and succeed.

Sensory specializations

Predatory animals provide some of the best examples of ultra-specialized sensory equipment, probably because their way of life demands a high degree of efficiency in hunting. One way for them to enhance their effectiveness is to increase the range over which a prey can be located. Another is to be able to pinpoint the position of a prey animal even in conditions which would normally provide protection. Both these methods involve refinements of basic sensory organs such as ears, eyes and nostrils or the construction of completely novel equipment.

Vision

Man is a visual animal; most of the detailed analysis of the world around him comes via his eyes. Consequently, he can comprehend the adaptations of predator vision that have occurred in other animal groups, but his own perceptions fail when it is a question of such exotic abilities as gaining information from electric currents, ultrasound or infrared radiation.

Many predators' eyes are more sensitive than man's to fine details of the landscape. For example, certain predators are able to see a mouse moving in grass at a range of half a mile; to man's eyes, however, the mouse would have merged into the background at that range. Daytime hunting birds of prey such as eagles, buzzards, harriers, falcons and vultures usually find their prey or carrion visually and are able to do so with precision over long distances. Their eyes are relatively much bigger than man's, and their retinas have

dense concentrations of especially small, light-sensitive cells which are responsible for colour discrimination.

Large eyes are also found in a number of nighttime predators such as bush babies searching for insects among African trees, owls hunting for shrews, mice and voles, and nightjars hawking for nighttime moths. Their eyes capture any available light and quickly and completely adapt to dark conditions. Their retinas have many rod light receptors that work best in low light conditions and provide only a black and white perception. Similarly adapted eyes are found in deep-sea predatory fish and some cave-dwelling, insect-eating salamanders. No predator, however, can see in complete darkness; in that situation other senses have to be utilized.

Hearing

The sounds emitted by prey animals betray their position. Several groups of predators have particularly large outer ears to gather the tiniest sound made by prey. One of the nocturnal hunting foxes, the tiny fennec of the Sahara, has huge ears that function as sound-receiving aerials and as radiators to get rid of excess body heat in the murderous daytime temperatures of the desert.

Other subtleties are apparent in predatory hearing; owls can locate a living mouse in complete darkness by extremely acute hearing that provides an accurate directional fix on the point of sound emission. Fish, using their internal ears and their lateral line system, can pick up high frequency sounds and water disturbances. In bony fish, the lateral line is a channel in the skin which communicates, through pores, with the water outside. Water disturbances enter through the pores and disturb vane-like sensory flaps, neuromast organs, set in the sides of the channel. The African clawed toad, which hunts in cloudy water for insects and crustaceans, has similar neuromast organs set externally in the skin of its head and body.

Almost certainly the predators with the most sensitive hearing are bats. These flying mammals most commonly hunt airborne insects at night. They can locate and intercept a midge or a moth in utter darkness by ultrasound radar. Bats emit bleeps of sound of a frequency too high for the human ear to hear. Echoes bouncing back from the prey are picked up by the bat's intricate ears. One type of bleep, the single frequency whistle, gives the bats information about the speed of the prey. A different bleep, which drops down through a scale of tones, enables the position of the prey to be pinpointed accurately.

Smell and taste

In man these two sensory abilities are separate. In many animals, particularly aquatic ones, the distinction between the two senses is blurred. Basically, both senses relate to the perception of chemicals in the environment—those which may be picked up and applied to an organ of taste.

Many aquatic predators appear to be able to use such chemical senses to home in on prey targets from considerable distances. Sharks show this ability to a marked degree. They seem to respond to substances from fish skin mucus that dissolve in water currents and, as a result, are able to track down the fish that produce them. It has been suggested that the bizarre head of the hammerhead shark, with its lateral wing-shaped extensions, is an adaptation for especially sensitive chemical hunting. The external nostril openings are near the tips of the wings and thus the two intakes are exceedingly far apart. This wide spread probably allows the shark to obtain a more accurate fix on the source of any attractive food.

Certain terrestial predators such as snakes also have a refined set of chemical senses. When a snake extends its forked tongue and then retracts it into its mouth, it is testing its immediate environment for information-giving chemicals. These are picked up on the

Fangs, teeth and beaks are all weapons used by a predator when attacking or eating prey. Snakes use their poison fangs to incapacitate their victims. The powerful teeth of dogs and hyenas are used both to kill and to gnaw the flesh of prey, and the eagle uses its hooked beak to tear its prey apart for eating.

Snake

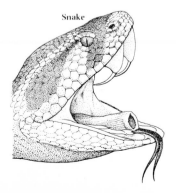

Hyena

Eagle

tongue and applied to a sensory pit in the roof of its mouth. The trails of prey animals can be sensed and followed in this way.

The high-precision tuning of ordinary sensory systems is only one side of the hunters' repertoire. Specialized carnivores have evolved completely novel means of picking up information concerning their prey.

These predators have developed vibration-sensitive systems both under water and in the air. Spiders' legs carry vibration receptors that enable them to interpret the movements of their web. They can probably tell the difference, simply by the type of vibrations caused, between the wind blowing the web, the arrival of an edible fly, the more violent struggles of a dangerous bee that must be cut free, and a harmless falling leaf. The anterior ends of marine arrow worms carry vibration receptors as well as vicious spiny jaws. These sense cells enable the tiny spear-shaped predators to accurately chase animal plankton whose tails or limbs produce the tell-tale water movements.

Some snakes have extra "eyes" that respond not to ordinary visible light waves but to the infrared radiation of radiant heat. At night, a snake equipped with this infrared homing device can pick out the position of a warm-blooded mammal or bird that is warmer than its surroundings. Many so-called "heat-seeking" air-to-air missiles home in on the hot exhausts of enemy planes in much the same way.

It is only in the last twenty years that the most extraordinary set of senses in a group of predators has been discovered—the electric senses. Some animals, especially fish, can respond directly to minute electrical currents. Sharks use this almost magical ability to find prey fish that may be hidden under a concealing layer of sand. Even a motionless flatfish generates enough electrical activity in its nervous system and muscles to be picked up by the cruising hungry shark.

Certain African fresh-water fish such as *Gymnarchus* live in muddy, cloudy water in swamps and slow-flowing rivers and use an electric sense for navigation and hunting. Each fish has a field-generating organ that sets up a weak electric field in the water around it. Any object that enters this field causes distortions, which can be picked up and analyzed by the fish and the information used to avoid obstacles or choose prey animals.

Adaptations for killing

In terms of predators, nature is often said to be "red in tooth and claw". This description is an apt one because most vertebrate hunters use modified teeth and specialized claws on the feet to subdue and kill prey. The basic killing implement is usually a hard-pointed, impaling weapon. Teeth, for example, are capped with enamel, and in many groups can be replaced, or are constantly growing.

The functionally similar claws are made of keratin, the same horny substance from which reptile scales, mammalian hair and bird feathers are constructed. Predatory reptiles, such as crocodiles, have huge death-dealing mouths full of conical teeth. Most predatory mammals have sharp claws and pointed canine teeth near the front of the mouth, which are longer than their neighbours.

Birds of prey have powerful hooked beaks for tearing at their prey, but their main killing weapons are the four-toed talons. Each toe is tipped with an inward curving claw so that as the foot closes on the doomed prey, it is impaled from four directions at once. A few predatory birds have modified the horny parts of their beaks to produce what is effectively a toothed bill. In doing so, they are regaining a function lost by the reptilian ancestors of the birds over 100 million years ago. The best-known examples of birds that have turned back the evolutionary clock in this way are the fish-eating diving ducks such as the

goosander, merganser and smew. They need the "teeth" to hold their slippery prey under water.

Other vertebrates have developed more exotic killing methods. Perhaps the most extraordinary predatory technique is the stunning or killing method used by the electric catfish in African rivers, the electric eel (rejoicing in the name *Electrophorus electricus*) of the Amazon and the electric rays of the Atlantic and Mediterranean. These fish are able to deliver massive electric discharges into surrounding water. The electric eel, for instance, can produce a shock of 550 volts, which it uses to kill fish and frogs that come near it. The dead or stunned prey is then eaten.

The legless snakes, without the possibility of modified limbs or feet for attacking weapons, have developed a range of potent methods for incapacitating their prey. Boas constrict and crush their prey in the muscular coils of their long body before swallowing them. Many other snakes such as vipers, cobras and rattlesnakes use poison. They inject paralyzing or tissue-dissolving toxins into their prey with hollow fangs set in the upper jaw.

Chemical warfare is common among invertebrate animals, too. Corals and jellyfish possess microscopic stinging cells, each one capable of injecting toxins into a small fish. Spiders have hypodermic needle-like fangs set in appendages, which they bury in their prey to inject poisons. Scorpions have a muscular poison bulb equipped with a hollow sting at the end of their tails, and a wide range of wasps and bees carry a toxin-delivering sting in this same "rear gunner" position.

Most invertebrates, however, kill by the use of mouthparts or limbs. The beak of an octopus, the jaws of a ragworm and the poison tooth of the deadly coral reef cone shell are all examples of the first tactic. The suckered arms of a squid or the fearsome, jagged, anterior legs of the praying mantis are horribly efficient examples of the second.

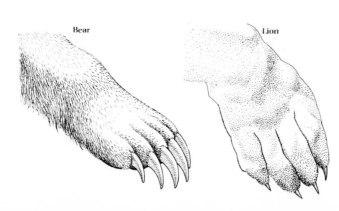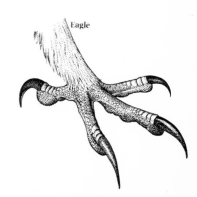

Claws and talons of all kinds are powerful killing weapons. The bear's long, non-retractile claws are ideal both for attacking large prey and for finding other food such as plants and fish. The lion's retractile claws enable it to grasp and bring down its prey; the eagle seizes its victim and crushes it with its strong talons.

Bear

Lion

Eagle

The Underhand Predators

The lion and the eagle—powerful, aggressive carnivores that grapple with their prey in a face-to-face battle, then devour the dead victim whole and at once—are the stereotype hunters. But there are other animals that have less ostentatious, yet perfectly efficient ways of obtaining a live animal diet. From fleas and tapeworms to whales and flamingoes, all these animals, with their different methods of feeding, are vital links in complex interconnecting food webs.

Netting a meal

Many animals have come to the evolutionary conclusion that it can be just as easy to eat a million tiny animals in a day as to catch a single large one, as long as the minute creatures are sufficiently numerous or can be concentrated together quickly enough. These

afterwards ready for the next meal.

Variations on the filter feeding theme abound. The sponges, creatures so simple that they do not even have a nervous system, are all filter feeders. Indeed there is little more to a sponge than its filtering system. A sponge is built round a complex of water-conducting tubes. Sea water is taken in through myriad minute holes all over the sponge's surface and eventually passed out through one or more large holes, the food particles having been extracted.

Water is moved along inside the sponge by the cells lining the tubes. Each of these cells is equipped with a thin, mobile projection known as a flagellum, and it is the massed beating of thousands of these flagella that moves the water. Every flagellated cell also has its own minute filter which fits round the base of the lash like a collar. Bacteria

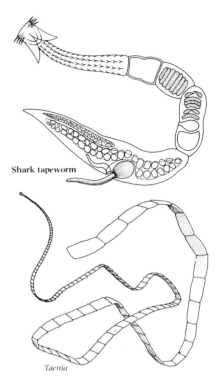

Shark tapeworm

Taenia

Tapeworms are gutless parasites which live in the intestines of vertebrate animals and absorb food across their body walls. The specialized tapeworm, top, from a shark-type fish, is made up of only a few segments but *Taenia*, above, from a human gut, can be many feet long and have thousands of segments.

Mucus bag

Chaetopterus

The bizarre annelid worm, *Chaetopterus*, builds a "U" shaped tube on the sea bed. Muscular, fan-like parts of the body waft water through the tube and wing-like arms support a mucus bag which acts as a food-catching net. The bag and food are swallowed.

Mucus bag

Sponge

Water, containing food particles, enters through tiny pores on the outer surface of the sponge. Internal "collar" cells each have a flagellum to pull water into the pores, and cells to ingest food. Water flows out through larger openings on the inner surface.

Outer surface pores

Living on others

One group of predators has perfected the trick of living on or in other animals and eating them so slowly that their "prey", known as hosts, can usually replace lost parts at the same rate as they are removed. Such predators are called parasites, their way of life parasitism, which is really an infinitely long predation in which the host animal dies only rarely.

This life-style of "underhand predation" has advantages. A parasite living inside its host resides in such a well-stocked larder that shortage of food scarcely ever limits its growth or development. But to be successful the parasite, which can usually survive only in a living host, must surmount a number of problems. Feeding rates must be carefully controlled to keep the food-providing host alive—too many parasites in one host, the removal of too much host material in a short time, or damage to a vital host organ, all threaten death of the host and with it the inevitable death of the parasite. The successful parasite must also be able to resist the host's attempts to remove or dislodge it, and its offspring must make landfall on another suitable host if they are to survive. The chances of finding

filter feeders, as they are known, extract their food from water. By straining or filtering vast quantities of it they can consume enough food to sustain bodies far larger than those of their prey. The blue whale, the largest animal that has ever lived, is tribute to the success of filter feeding. It exists on vast quantities of krill, animals like shrimps, each only an inch or so long.

All filter feeders need a "net" in which to catch their food. Sometimes the net is exactly that—a fibrous bag through which water is strained. The marine worm *Chaetopterus* feeds in this way. It lives in a tube buried in mud or sand and has paddles to draw water through its filtering bag. When the bag is full of plankton *Chaetopterus* eats the meal, bag and all, and secretes a replacement bag

and algae in the water stream are trapped by these filters, then engulfed by the cells and digested.

Flamingoes, bizarre pink birds found mainly in the saline lakes of Africa's Great Rift Valley, are specialist filter feeders. The greater flamingo, *Phoenicopterus ruber*, filters fine mud through its ungainly bill fitted with sieve plates and extracts from it the tiny animals it feeds on such as crustaceans, worms and larvae.

Filter feeding may seem laborious but it does have some points in its favour. While an attacking lion may be fatally wounded, for example, by the sable antelope that it assaults, there is no chance of krill ganging up on a whale or shrimps mounting raids on ducks or flamingoes.

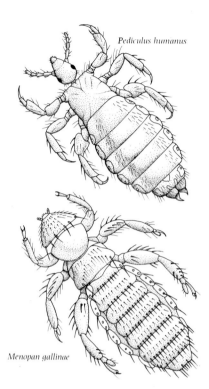

Pediculus humanus

Menopan gallinae

The human body louse, *Pediculus humanus*, feeds on blood by pushing its mouthparts into human skin to suck from superficial blood vessels. The feather louse, *Menopan gallinae*, lives on birds and feeds on feather protein. The feathers are chewed, then ground up in the gut before digestion.

exactly the right host are often small.

For centuries the parasitic human louse, *Pediculus*, has made its home on man. These wingless insects spend their lives on human skin, clinging on doggedly to skin or hairs with hooks on the ends of their legs. If the lice fall off their host and do not attach themselves immediately to another person's skin they are dead.

Adult female lice lay eggs (nits) which they stick on to the hair with quick-hardening cement. Both the young that hatch from these eggs and the adults feed on human blood which they suck up using hollow, piercing mouth-parts. The minute amounts of blood taken by even a large population of lice are made up almost as soon as they are removed. Blood is a highly nutritious diet, but is low in B vitamins. To overcome this, the lice themselves play host to helpful bacteria which, in return for a cosy refuge, provide essential B vitamins for the lice.

The reputation of model parasite is held, deservedly, by the tapeworm; it is one of the most highly adapted internal parasites. Tapeworms evolved from free-living flatworms which, hundreds of millions of years ago, probably hunted

for live prey in salt and fresh water, using an extendible muscular tube from the mouth, and digested it in a gut. Today's adult tapeworms live a much altered life. They have no gut at all. Instead they live inside the small intestine of birds, mammals, fish and other animals and absorb the host's digested food across their body walls.

The constant movement of the host's intestine makes it a precarious home, but tapeworms have evolved a solution in the form of a collection of hooks, suckers and clamps at the front of their bodies which provide firm anchorage to the host's gut wall.

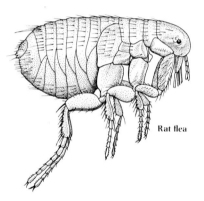

Rat flea

Fleas live in their host's nest or home and jump up to feed on the host animal. The rat flea carries bubonic plague bacterium.

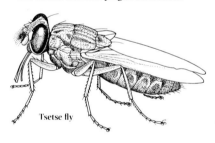

Tsetse fly

The tsetse fly is harmful to man in two ways. It is a blood feeder using its probe to bore down to blood vessels, at the same time injecting anticoagulant saliva. In this way it transmits disease to man. Its salivary glands may contain infective stages of the trypanosomes that cause sleeping sickness and these are injected as the tsetse feeds. If the host is already infected he infects the fly.

Using the unlimited food supply that swirls past them day after day, tapeworms produce hundreds of thousands of eggs. This stupendous reproductive output, which is essential to offset the astronomic unlikelihood of the offspring ever finding their way back to a correct host, is achieved by turning themselves into something shaped like a pearl necklace. Each "pearl"—there may be as many as 10,000 to a worm—is a segment containing complete male and female reproductive systems.

The insect killers

Unlike parasites, which rarely commit suicide by killing their hosts, parasitoids, or insect parasites, always kill their hosts and always survive the host's death.

Almost all parasitoids are two-winged flies, ants, bees or wasps. The hosts that they use for food are other insect species, spiders and wood lice. The pupae, caterpillars and other developmental stages are often the targets for attack. Usually, the adult female parasi-

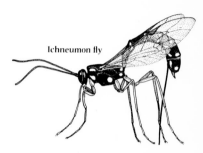

Ichneumon fly

The ichneumon fly, *Rhyssa*, is a parasitoid closely related to bees, wasps and ants. The female inserts her eggs into the bodies of other insect larvae. When the *Rhyssa* larvae hatch they feed on the internal tissues of the host larvae.

Chalcid wasp

Chalcid wasps are tiny specialist parasitoids belonging to the hymenopteran order of insects. The female of the chalcid *Pteromalus puparum* locates caterpillars of the cabbage white butterfly which have become pupae—the immobile stage between the larva and adult—and lays her eggs inside them. When the *Pteromalus* larvae hatch they feed on the butterfly pupae.

toid lays her eggs in or near the host species. The larvae that develop from these eggs grow inside the host, consuming it slowly from within.

Far from being obscure or inconsequential, this way of life has great ecological importance. The insects that obtain their food in this way, and there are probably 100,000 different species of them, play a crucial role in the natural control of insect populations. This has prompted man to use them artificially to regulate insect pests.

A Vital Chain of Feeding Links

How does the predatory, carnivorous animal without any obvious productive contribution fit into the pattern of life styles that exists in all natural communities? The answers to this question are complex, and are to be found in ecological research that examines overall energy flow through ecosystems. The term ecosystem describes a community of organisms, both plants and animals, and the particular environment in which they live, such as a river, wood or jungle. All the organisms in the community interact with one another and are influenced by the conditions of their environment, such as the temperature and vegetation.

The energy needed to maintain the living activities of the animals and plants in an ecosystem first enters the community of living things as light energy from the sun. By the absolutely fundamental process of photosynthesis that occurs in green plants, some of this light energy is converted into chemical energy in the form of organic substances such as sugars. During photosynthesis, which is centred on the chlorophyll and enzyme systems in chloroplasts, carbon dioxide from the air and water from the soil or the aquatic environment are fused together to form glucose. Such synthesized substances in green plants are a primary store of energy that must power the activities of the plant. Using other elements such as nitrogen, sulphur and phosphorus, obtained in the form of dissolved mineral salts, a plant

can expend part of its energy store to construct vital substances such as proteins, fats and DNA.

Green plants are independent feeders in the sense that they do not depend on the activities of other organisms to provide their food. However, for all the other organisms in the ecosystem, principally animals, the story of energy flow has just begun. The massive energy store represented by the synthetic abilities of plants powers the activities of all the animals that share the plant's environment. The production of organic substances in the plants of an ecosystem is termed "primary production". Although specialized carnivorous predators rarely, if ever, eat any of this primary production directly, they are absolutely dependent on it. For example, the wildebeest would not be around for a lion to eat if grass were not growing in that same habitat and trapping energy from the sun.

To come to grips fully with the essential features of the predator's way of life, the ways in which energy flows through ecosystems must be understood. The pattern of who eats whom in any community of animals and plants, such as the organisms in a lake or the mixture of organisms in the plains of East Africa, is intricate. But there is an overall structure that tends to recur in many different ecosystems. First, there is a group of microscopic (algal) and large plants known as the primary producers. Herbivorous animals feed directly upon

them in a variety of ways: they can take pollen or nectar; or eat fruit or seeds; or graze on crops like grasses; or browse on larger bushes and trees. Many of these animals have adaptations that enable them to grind up the fibrous and cellulose-containing parts of plant food as well as to digest resistant substances such as cellulose itself, which is the commonest organic chemical in the world. These herbivorous animals are termed primary consumers because they are the first animals in a chain to make use of the energy locked up in green plants.

The primary producers and the primary consumers represent successive steps in the criss-cross system of feeding links between different species known as a food chain or web. Each of these steps is called a trophic level. All the subsequent trophic levels relate to predatory animals.

The predatory animals that feed directly on the herbivores are called secondary consumers, and they are a varied group. They must be, of necessity, because the herbivores that they feed on are themselves fantastically diverse. Some of these predatory animals are adapted to feed on insects that either feed on nectar and pollen—bees, wasps, butterflies, moths—or plant sap—aphids—or bulk plant material—beetles, ants, termites. Within this category are birds such as warblers, tits, bee-eaters and honey buzzards and mammals such as insectivorous bats,

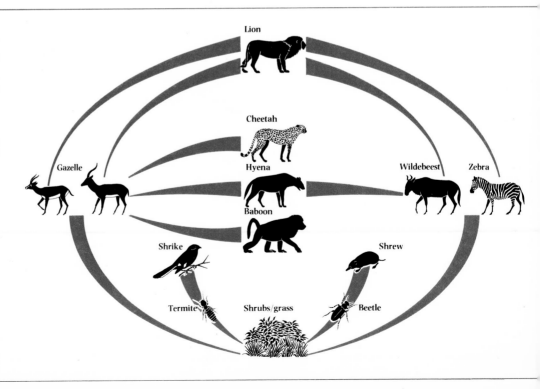

In the East African plains ecosystem the pattern of feeding links is complex with many different animals at each feeding level—herbivores and various grades of carnivores. Some of the most important feeding links in this intricate web are illustrated, right. The lion, for example, feeds on large herbivores such as zebra or wildebeest but, in its position as top predator of this ecosystem, can feed on almost anything else available. The shrike and the shrew are small insectivorous predators, themselves vulnerable to predation by larger animals and birds. Everything ultimately depends on primary production—the shrubs, grasses and plants.

Lion

Cheetah

Gazelle

Hyena

Wildebeest   Zebra

Baboon

Shrike   Shrew

Termite   Shrubs/grass   Beetle

anteaters, pangolins and aardvarks.

In addition, much larger herbivores exist, and predators (secondary consumers), which are even larger, feed on them. In Africa, for example, Thomson's gazelle crops grasses and browses on acacia bushes and, in turn, is preyed upon by a number of predators such as cheetahs, hunting dogs and lions.

Following the chain of feeding links even higher, a complex zone consisting of the carnivores that eat other carnivores is reached. In the system in which a pike eats a trout that has fed on water snails that have browsed on water weed that has trapped sunlight energy, four trophic levels are present. The weed is the primary producer, the snails the primary consumer, the trout the secondary consumer and the pike the so-called tertiary consumer. The pike in this particular lake, if nothing fed on it, could also be called a top predator.

Unfortunately for man's organizing mind, the predatory part of the stack of trophic levels is rarely so neatly layered. Many higher carnivores do not confine their attentions solely to the trophic level immediately beneath them. Several carnivores eat fruit and other plant material when their normal prey is not readily available. Foxes, for instance, will eat nuts and berries in hard times. Likewise, a bird such as an owl will readily capture and eat herbivorous field mice and insectivorous shrews. The owl does not concern itself with trophic levels; from its point of view

the shrew and mouse are available prey species that require similar effort to be captured. In fact, the higher the trophic level of a predatory animal, the more likely it is that its selection of prey species will spread across several lower levels. This phenomenon, taken to its extreme, produces a category of animals that may be termed omnivores or opportunist feeders. An excellent example is the Kodiak bear, which feeds on berries, honey, small mammals, birds and salmon.

Not all organisms are eaten alive; many die from infections or climatic accidents such as a sudden very cold night. Their bodies are consumed by a wide variety of organisms often referred to as decomposers. Some protozoa and many bacteria are in this category. Strictly speaking, any carrion eater—hyenas, vultures and sometimes even the lion—must be regarded as a decomposer.

The flow of energy through the trophic levels, however, has other important implications for our understanding of the place of predators in the natural world. The most crucial is the constraint imposed by the efficiency of energy transfer between levels.

In any food link where one species eats another, there is always an energy loss. For example, when blackbirds feed on beetles, the energy that should enter the blackbird population as beetle food gets lost before it can be turned into more or heavier blackbirds. First, only a

part of the beetle can be digested—hard chitinous parts of the skeleton cannot be broken down in the blackbird's gut and, as a result, they pass out in the faeces. Of the digested food that is absorbed, a large proportion will be used as "fuel" by the birds to perform work, such as moving about to catch more food, or is simply lost as heat. The work energy and the heat and respiratory losses are completely lost to the blackbird population. Organic substances with significant energy content are also lost in excretion and at the time of death. For any consumers of blackbirds, say a cat population, only a small proportion of the beetle energy that went into the blackbirds is now available in the next trophic level.

These inevitable energy losses are cumulative through the trophic levels. Farther and farther away from the green plant primary producers, the total quantity of accessible energy in each trophic level gets less and less. In addition, predators are usually bigger than their prey. These two facts mean that numbers of animals in each trophic level of an ecosystem decline from primary consumers to the top predators. Although omnivorous species complicate the calculation, it will normally be true, for instance, that in an area of woodland containing a pair of owls there will be hundreds or thousands of small mammals such as rodents and many million insects—a living pyramid of animal numbers.

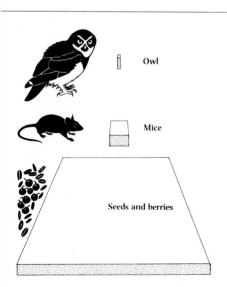

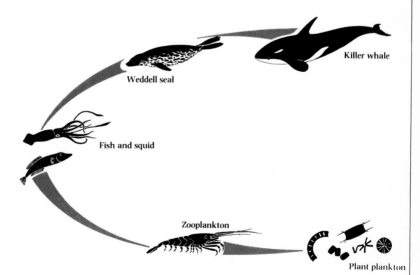

In the sequence of an owl eating small herbivorous rodents, a pyramid of both total biological weight (biomass) and species numbers results. For each owl there will be hundreds of rodents and an even greater mass of seeds and berries.

Microscopic floating plants, phytoplankton, are the primary producers in the sea, as green plants are on land, and the basis of all marine food chains. Plant plankton is consumed by zooplankton, tiny, passively-drifting animals such as the

shrimplike krill. Many fish and invertebrates feed on the vast numbers of planktonic animals; they in turn are preyed on by large marine creatures such as seals. Topping the chain is the mighty killer whale which will eat seals.

The Geography of Life and Death

The predators which live astride the food webs of the world, making a living at the expense of other animals in the intricate chain of existence, differ from continent to continent. But although the actual species found in the food webs may be dissimilar, in all the various regions of the earth there are predators and prey playing similar roles in the scenario of life and death.

Like the blocks of different shapes designed to fit the holes of a child's "mail box" toy, most of the species within any food web are adapted to survive in a particular niche or section of the environment. And as every hole in the mail box is limited to a particular shape, so every niche is restricted in its living conditions, including the food that it provides.

The history of life on earth reveals that if a potential way of earning a living exists in an environment then some organism will evolve to fulfil that potential. Isolated oceanic islands are excellent examples of the way this works. As if by a lucky accident, many of these islands are ready-made "natural laboratories" for studies in evolution. On a continent such as Europe the number of animals competing for survival is huge, but only a few animal groups may have completed the long and perilous journey to an island that is difficult to colonize. Once they have arrived, however, these island colonists are presented with a wide range of possible niches into which they can expand. As a result, a single group of animals may diversify and evolve into the specialists whose roles on a continent would be shared among quite separate animal groups.

The world's most famous living laboratory is the Galapagos, a group of volcanic islands in the Pacific. Charles Darwin visited the Galapagos in 1835 during his voyage on HMS *Beagle*, and the patterns of animal life he saw there helped to shape his epoch-making ideas on natural selection and evolution. One group of birds, the Galapagos finches of the family Geospizinae, dominates the avian life of the islands. These finches must have been among the pioneer land birds flying in from South America after the Galapagos islands rose out of the sea. An almost complete lack of competition meant that the finches could fill many different niches virtually unhindered, and today the original colonists, which were seed eaters, have evolved into 14 distinct species, none of which can inter-breed.

Of these new species, some still eat seeds. Among them are species with gigantic beaks that are used to attack very large seeds and berries and a variety of forms with lighter, narrower beaks, which feed on smaller seeds. Other finches have, however, taken on life-styles which, in a more competitive situation, would be the province of quite separate bird families. Those called Certhidae, for example, have developed into thin-billed, warbler-like insect eaters. The remarkable species *Camarhynchus pallidus* has developed an amazing predatory life-style. Creeping over the surface of cacti and bushes like a nuthatch, the finch finds a cactus spine, holds this tool in its beak and uses the spine, rather as a woodpecker would use its long tongue, to winkle out prey— insects, spiders and larvae—from deep holes and crevices.

On a much larger scale the whole Australasian continent reveals a similar pattern of niche-filling by a group of animals whose ancestry is restricted. Because Australia drifted apart from the other continents when pouched marsupial mammals dominated the world, these creatures were never exposed to the full competitive force of more modern, placental mammals. So in Australia pouched mammals play the predatory roles taken by quite a different cast of animals in other regions of the world—the Tasmanian wolf, for example, is a dog-like carnivore and the Tasmanian devil and tiger cat are, like the mongoose, generalist predators. The pouched, ant-eating wombat, with its long snout, narrow tongue and powerful digging claws, fills just the same ecological niche as the anteaters of South America or the aardvark in Africa. There is even a burrowing marsupial mole that eats worms and insects from under ground in exactly the same way as the European mole.

When the jigsaw of animal distribution is finally pieced together it is possible to see that different, extensive regions of the world have animal families largely restricted in distribution to these regions alone. Zoogeographical regions or faunal realms are the names given to these expanses of the earth and each has its own characteristic groups of animals. The realms are on the scale of continents and seven of them are usually recognized: the Palearctic realm, or Old North, a huge region extending from Iceland to Japan and from Spitzbergen to Arabia; the Nearctic, or New North, the whole of North America; the Neotropical, which includes the narrow land mass of Central America, the Caribbean and the whole of South America; the Ethiopian realm, basically the continent of Africa but including Madagascar and part of the Arabian peninsula; the Oriental, formed by the Indian subcontinent, part of China and Southeast Asia and bounded by the Himalayas; the Australasian, which includes Australia, New Zealand, New Guinea and Sulawesi; and lastly the Antarctic, the most isolated realm of the seven.

While some groups of predators are not exclusive to a particular zoogeographical location, such as the eagles, herons, grebes, nightjars and cormorants which live in all realms but the Antarctic, many predators have a much more restricted distribution. The tyrant flycatchers, for example, which hunt and eat insects in the same way as shrikes, are a bird family found only in North and South America. Similarly, the spiny anteater, or echidna, lives only in Australia.

The mixed communities of plants and animals within each faunal realm impose a specific biological appearance on the landscape. Chief architects of this landscaping are the dominant species of plants, whose distribution is, in turn, moulded by the forces of nature. These include rainfall, temperature, altitude and soil type on land, and salinity, temperature and depth in water.

Large, stable, recognizable communities of plants and animals are known as biomes. On land there are three sorts of tree-dominated or forest biomes— coniferous, deciduous and tropical. Shorter plants such as grasses predominate in savanna and temperate grasslands, while the cold tundra and dry desert biomes support only a limited number of species because of the severe restraints imposed by the environment. Intermediate biomes such as chaparral and open woodland also exist. In the sea up to four communities are present, depending on the depth of the water and the nature of the sea bed.

The pattern of vegetation within a biome is the loom on which the fabric of animal life is woven, for the plants not only form the physical landscape but also determine the range of herbivorous life-styles which are possible. And with these two features fixed, the pattern of predatory life is, to a large degree, predetermined. The predatory animals within any biome do, however, show finely tuned specializations and adaptations to life in those specific surroundings. This is excellently illustrated by the predators which inhabit coniferous and tropical forests.

In the intense cold of the far north of the Nearctic realm of North America is a vast belt of coniferous forest stretching from one side of Canada to the other. The trees that grow in the infertile, acid soil are small but densely packed together, with spruces, firs and pines predominating. Compared with the gentler environments farther south, the

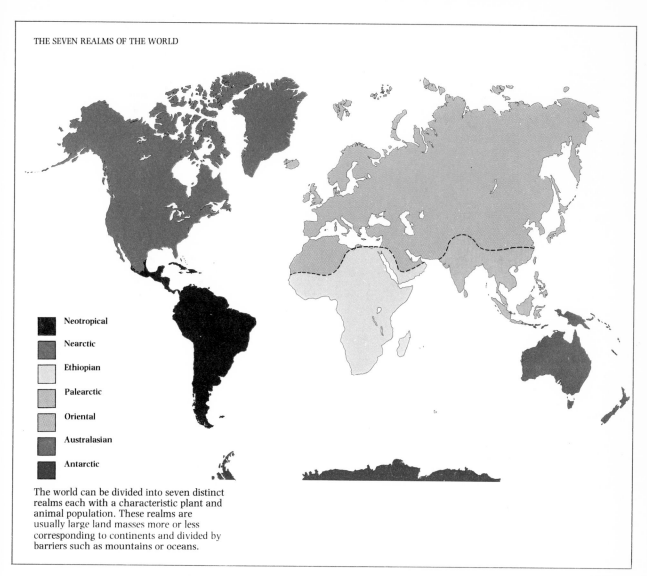

- Neotropical
- Nearctic
- Ethiopian
- Palearctic
- Oriental
- Australasian
- Antarctic

The world can be divided into seven distinct realms each with a characteristic plant and animal population. These realms are usually large land masses more or less corresponding to continents and divided by barriers such as mountains or oceans.

number of tree species is small, but the actual number of trees huge. In more open areas, a short, tough vegetation of grasses and shrubs carpets the ground below the trees.

The herbivores of the forest all depend on the coniferous tree cover for survival. Chipmunks, spruce mice and other animals feed on the seeds in pine cones. Birds use the same food source and specialists like crossbills and grosbeaks subsist almost entirely on this diet. An enormous range of insects feeds on the needles, bark, wood and roots of the trees and croppers of low vegetation such as the porcupine and moose often take to bark stripping during the hard winter months.

An interesting mix of predators is generated by this selection of herbivores, given the constraints imposed by the severe cold and extreme seasonability of the climate. Pack-hunting wolves foray over wide areas of the forest. Minks will swim to catch fish, burrow for small mammals or climb

trees to search out eggs and nestlings. Among the forest trees, the great horned owl swoops down to grasp any available small mammal with its powerful, pointed talons.

In sharp contrast to the coniferous forests, the tropical rain forests are enormously complex biological systems containing many plant species. The bushes and trees of the forest reach skywards in a series of vertical layers of foliage to a tree-top canopy layer about 75–100 feet (23–30 metres) above ground. Individual trees can reach to over twice this height. The great diversity of tree species, plus the vertical layering of the plant environment, provides the animals with many ecological niches and as a result a huge variety of herbivorous and carnivorous animals flourishes. This variety has been most closely studied in the tropical forests of the Amazon and those of central Africa.

Among the predators of the tropical forests is a staggering selection of

insect-eating reptiles, amphibians, birds and mammals. Chameleons, tree frogs, marmosets, anteaters and flycatchers—and their many insectivorous relatives—thrive by taking advantage of the teeming insect life in these warm, humid jungles. Snakes, excellently adapted to life on the forest floor, feed on a variety of other reptiles, amphibians, mammals and birds.

In the close-packed vegetation, large mammalian predators are rare beasts. Instead there are many small insectivorous mammals such as shrews and generalized small predators like the agile forest genet. In both South America and Africa, many of the monkeys and apes live omnivorous lives. All tropical forests also seem to have encouraged the evolution of massive daytime hunting birds of prey that attack larger animals in the high canopy. The harpy eagle of the Amazonian forest catches and eats arboreal mammals, while in the Philippines the monkey-eating eagle feeds exactly as its name suggests.

Predators and Prey: the Numbers Game

The world's populations of plants and animals are in a state of constant change. The investigation of these communities, including the way they alter from year to year and season to season, the way they differ from place to place and the environmental influences which mould these patterns of change, is the essence of the science of population dynamics.

The factors that determine exactly how the numbers of individuals in an animal population fluctuate are both physical and biological. Temperature, rainfall, soil and the configuration of the landscape—for example, the number of caves in an area—are among the physical conditions that may have a powerful influence on the size of a population. Biologically, population sizes are modified by the ways in which different species of animals interact. The two most fundamental of these interactions are competition and predation.. All animals compete for the resources their environment provides and there are few animals which do not suffer from some kind of predation.

Much scientific attention has been focused on the population dynamics of predatory animals and their prey species. The reasons for this are chiefly practical. If sensible management and conservation policies are to be achieved in the ever-shrinking areas of the earth where an attempt is made to maintain mixtures of animals in near-natural

surroundings, then the effects of predator and prey populations on one another must be meticulously worked out. In a North American national park, for instance, it is crucial to be able to predict the influence that different concentrations of wolves might have on deer numbers.

Man can use the finely tuned interactions between predators and prey to good advantage in the natural control of the diseases and pests that afflict him, his domesticated animals and his crops. Predators and parasitoids can often be used as agents in biological control programmes to augment or even replace the use of drugs or chemicals.

Adult mosquitoes are the vectors whose bites spread the potentially fatal diseases of malaria and elephantiasis. But rapidly-breeding insectivorous fish such as *Gambusia* can be employed to help combat these diseases, for they feed on aquatic mosquito larvae and thus reduce the mosquito population. Similarly ladybird beetles can be effective natural control agents, for their favourite diet includes the aphids and scale insects that decimate crops. In Canada the winter moth defoliated huge numbers of hardwood and orchard trees before being successfully brought to heel by the planned introduction of death-dealing parasitoid flies and wasps into their community.

The evolution of new characteristics in both predators and prey has a marked

effect on the ways in which predator and prey populations influence one another. In prey animals, the laws of natural selection tend to legislate in favour of those specializations which lessen the chances of capture. It is this "selection pressure" which has brought about the amazing range of specific behaviour patterns, defensive organs, social defences, camouflage and mimicry that help prey animals evade predators. On the other side of the coin, natural selection will also favour the survival of any predator with increased hunting and killing ability.

These contrary evolutionary processes assisting both predators and prey continue unabated and side by side. They are the moves, counter-moves and counter-counter-moves in a gigantic biological chess game.

The success and growth rates of populations of prey animals are the chief arbiters of the rates at which predator populations increase or decrease in numbers. This must be so because the prey animals inevitably provide the bulk of the food necessary to ensure the survival of their predators.

The interaction between lemmings and the birds that devour them is one of the best-studied examples of the "numbers game" played between predators and prey. For a variety of complex reasons lemmings undergo violent population explosions and collapses. The difference in the density of lem-

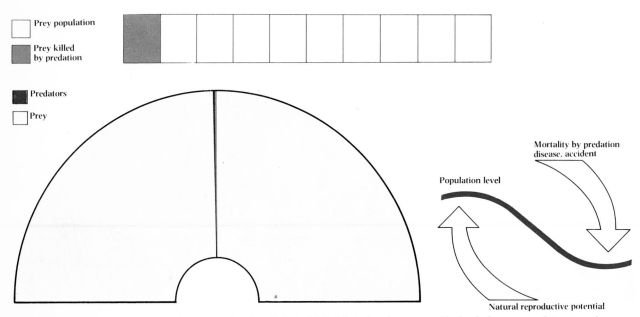

Prey population

Prey killed by predation

Predators

Prey

Mortality by predation disease, accident

Population level

Natural reproductive potential

Prey animal populations must always be greater than predator populations if predation is not to have too great an impact on prey numbers. According to surveys carried out in wildlife reserves in Africa there is as little as 2.2 lb (1 kg) of predator to 550–660 lb (250–300 kg) of prey in the Serengeti. In an average year, predators kill about 10 per cent of the total prey population at a rough estimate.

The level of a prey population is under continual pressure from death by predation, disease and malnutrition. Counteracting this is the upward pressure of the prey's reproductive potential.

mings in a particular part of the countryside can vary up to a thousand-fold over a period of about ten years.

At Point Barrow in Alaska, the birds that hunt and eat lemmings are the pomarine skua (a type of predatory seagull) and two owls, the magnificent white snowy owl and the short-eared owl. A rise in the lemming population was found to prompt a dramatic response in the population of these three predators, but each in a different way. When lemmings were rare, none of the birds bred at all. In the course of a single year, however, the breeding population of the skuas increased from four to 18 pairs per square mile as the lemming population climbed. The reason must have been that new individuals congregated in an area of abundant food. Through the lemmings' upturn in fortune, the density of breeding pairs of snowy owls stayed constant, but the clutch size of each pair increased enormously and many non-breeding snowy owls entered the area in search of food. The response of the short-eared owls was an increase in breeding pairs near the peak of the lemming population explosion.

Interdependence between the populations of predators and prey reaches its extremes when both undergo regular up and down oscillations in numbers. The system works like this. When many predators are present, the prey population becomes depressed and this change pulls the predator population down. Released from predation pressure, the prey population thrives, enabling the predators to increase in numbers again and so on.

It was once thought that predator-prey relationships interact in this way with the predators controlling—driving—the size of the prey population. More detailed analysis shows that this will only happen if the predator is the "key factor", causing most of the deaths that occur among the prey, and if the predator feeds on only one prey species. Nature rarely complies with these two conditions. Usually, parasitism, accidents, the effects of climate and the like, result in many more prey deaths than the activities of a particular predator. Equally, most predators can happily feed on several different prey species.

By this token, the lynx-snowshoe hare relationship, a much-quoted example of predators controlling prey numbers, is unlikely to be what it seems. Originally it was thought that the crashes in the hare population were a direct cause of the killing activities of the lynx. But almost certainly other factors, such as plant food supply, were providing the impetus to the hare cycles and the changes in the lynx population merely followed the hare numbers rather than driving them. The proof of this was verified when the population dynamics of the snowshoe hare were charted on Anticosti Island in the Gulf of St Lawrence. Here there are no lynxes or other large predators, but the hare population still goes up and down in just the same way.

"Prey density" has an important effect on the rate at which the predator captures and consumes its prey, whether applied to a lion hunting in the Serengeti or a praying mantis eating flies from a leaf. As the number of flies surrounding a praying mantis increases, for example, the mantis eats them at a rate which rises smoothly to a saturation level after which any further increase is impossible. This maximum eating rate is achieved more quickly if the predator is starved first.

The feeding pattern of vertebrate predators, such as lions killing and eating gazelles or deer mice eating sawfly cocoons, shows a distinct difference compared with that of insects. As prey numbers escalate the killing rate also rises, but it does not climb regularly. When prey are scarce the killing rate rises slowly. This lag probably occurs because the predator's "hunting computer" tells it that great effort spent on that prey would be wasted. When prey become more plentiful, however, the rate of kills increases hugely as it becomes more and more profitable for the predator to concentrate its exertions on the prey. Ultimately, though, a maximum rate of predation is reached.

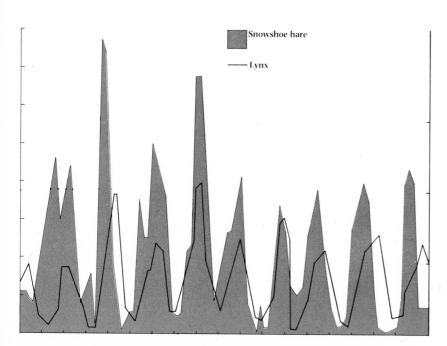

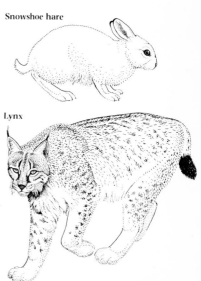

Snowshoe hare

Lynx

Hunters' fur returns in the area of the Mackenzie River in Canada provide the evidence for the interplay between the populations of the snowshoe hare and its predator the lynx. The fur records over a period of 90 years from 1845 to 1935, enabled the populations to be accurately charted. During this time, both bounced up and down dramatically with the peaks of each species being out of phase with one another. It seems that the hare population cycles are driven by changes in the climate and in the abundance of its plant food, while the lynx population is influenced by changes in the numbers of the hare.

Man the Hunter

Man is the top predator who became the top, top predator. He is the most complex of all predators and the ultimate social or group hunter. Just as his modern primate relatives, the baboons and chimpanzees, achieve success by intelligent group action during hunting, so man's success as a predator evolved from the cooperative hunting methods of clans or family groups, combined with an adaptable body plan.

The group of primates destined to be man's direct ancestors probably became distinct about ten million years ago. And although the exact time of man's emergence as a recognizable species is a bone of scientific contention, recent fossil findings in East Africa suggest that tool-using men existed between two and three million years ago.

What can be said without any controversy is that from two million to about ten thousand years ago, man's history was that of a "hunter-gatherer" who killed and ate wild animals as an important and continuing part of his diet and satisfied his vegetable needs by picking fruit, leaves and berries and digging for edible roots.

Early man had no agriculture in any accepted sense of the word. The remains of hunting implements used through the ages prove that man's hunter-gatherer life-style continued up until about ten thousand years ago. Only then do the beginnings of agriculture emerge, so man's history as a hunter is more than two hundred times longer than his past as a farmer.

Today only some 250,000 people are still hunters and gatherers—less than 0.003 per cent of mankind—and the proportion is constantly falling. Scattered throughout the world in tropical, temperate and Arctic zones, they include the Pygmies of the African rain forest, the Bushmen of the Kalahari Desert, the Australian Aborigines and the Eskimos. In their hunting techniques, which are often sophisticated, they use the natural materials their habitats supply. Apart from the dogs that are their hunting aids and companions they have no domesticated animals.

As a hunter-gatherer, man lives like other group predators such as the hunting dogs. And like other opportunist omnivores his body is unspecialized. Man has thrived because of his intelligence, social structure and powers of organization—abilities that allow him to take part in complex group hunting activities and to use tools. Physically man has kept all his options open. No aspect of his body make-up forces him into a particular life-style. His upright, two-legged stance, stereoscopic colour vision and manipulative hands, with thumbs that can touch every finger in turn, mean that he is an adaptable fighter and gatherer but not a specialist.

It is man's manual dexterity, used in concert with his intelligence, that has made it possible for him to become a tool user. A few animals do use tools in a rather desultory way, but man is an inveterate, persistent and complex tool inventor, producer and user. The history of tool-using is almost as long as the history of man himself—some East African fossil finds from over two million years ago reveal shaped stone tools which were probably used as hand axes.

Today's non-agricultural, non-technological hunting people use a wide variety of techniques and tools (weapons) to capture and kill the animals which are their prey. The ingenuity and complexity of these weapons and hunting methods are just as great as those needed to establish simple agriculture. So the hunter-gatherers are not, it seems, "less intelligent" than farmers, but operate under completely different circumstances—usually in habitats patently unsuited to agriculture.

To kill animals on the run or birds in flight a hunting man uses many kinds of weapons, including clubs, spears, bows and arrows, harpoons, slings and bolas (balls connected by strong cord and thrown at a quarry to entangle its limbs). Clubs of some sort are common to almost all hunting cultures and possibly evolved from the digging stick, the simplest implement of the hunter-gatherers. Clubs can be carved from hard wood and are used either for short-range throwing or for finishing off

Hunting-gathering man lives in perfect harmony with his habitat. By plucking and digging for plant food, and killing animals only when necessary, he maintains the balance of nature.

The tools and weapons of hunting man are witness to his skill and ingenuity. The Tlingit Indians of Canada are among the few hunting peoples to work metals. This copper-bladed dagger has a pommel of musk ox horn. An Alaskan Eskimo arrow may have a head carved from stone set into a walrus-rib shaft. In the complex toggle-headed Eskimo harpoon a barbed whale-rib head is mounted on a bone or ivory shaft. The Andaman islander's adze, used for canoe and weapon making, has a blade ground from a mollusc shell bound into a shaft cut from a tree fork. The S-shaped bows of the South Andamanese, often made of nutmeg wood, are strung with plant fibres.

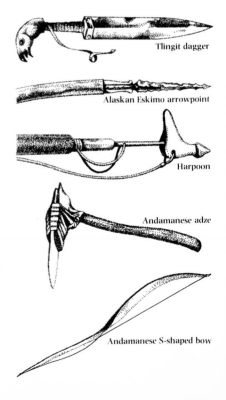

Tlingit dagger

Alaskan Eskimo arrowpoint

Harpoon

Andamanese adze

Andamanese S-shaped bow

wounded animals at close range. Light clubs can be thrown at flying birds and were probably the origins of the angled shape of the Aborigine boomerang.

Spears can be thrown farther than clubs and have sharp tips for piercing a prey's hide. Points can be fashioned from wood and fire-hardened, or made by joining pieces of stone, bone or shell to a wooden shaft. Range and power can be increased by the use of a notched throwing stick which effectively increases the hurler's arm length and hence the speed with which the spear can be thrown.

Bows and arrows are relatively modern hunting weapons. Man first invented and used them about ten thousand years ago—near the end of the last Ice Age. The mechanics of the bow mean that man's muscle energy can be stored in the weapon, then used quickly and violently when the arrow is released from the string, thus vastly increasing penetration power. In the pursuit of large, powerful game, the arrow (or blow dart) can be made even more deadly by smearing its top with poison. Both the Kalahari Bushmen and the Pygmies have made this kind of chemical warfare highly sophisticated by using mixtures of nerve and muscle toxins obtained from poisonous plants, fish and amphibians.

On the hunt, predatory man may employ his skills and tools indirectly. Tracking ability is an important asset.

sound-producing lures can call prey close to the hunter and disguises may help in approaching a quarry. The Maidu Indians of Central California hunt deer wearing deerskins and antlers. By cooperating together, a large number of hunters can capture enormous prey using only the simplest of weapons—the Eskimos of Alaska kill and bring home bowhead whales using only lances and harpoons attached to sealskin ropes.

The world's hunter-gatherers exist in equilibrium with their surroundings. Hunting-gathering man does not markedly alter the ecosystem he lives in—rather he is a natural component of it. As an efficient group predator and omnivore he does not wreak irreversible changes on his habitat.

Only farmers and industrial man have had the power to alter the face of the earth. They have replaced natural vegetation systems with their own domesticated crops and built over and destroyed the natural mix of animals and plants in those parts of the earth where their advanced technological capabilities have been unthinkingly brought to bear.

In modern industrial societies, the straightforward hunting of natural animal populations for food is relatively rare. Only the fish and whale schools of the world are extensively hunted to form part of the human diet. Man's ever larger appetite, his technological

expertise and his inability to restrict hunting levels to reasonable proportions means that this sort of hunting will disappear as species become extinct or reduced to minute populations.

Even more inexcusable is man's blatant defiance of the rules of nature in killing wild animals with motives of vanity and greed. "Fashionable" women's addiction to conspicuous display threatens the survival of all the world's spotted cats, particularly the leopard, jaguar and ocelot. The tiger, too, hunted not with primitive weapons but with sophisticated firearms fitted with telescopic sights, is diminishing in numbers yearly as it falls prey to the demands of sport and chic. And the unfortunate Javan rhinoceros, whose horns can supposedly act as an aphrodisiac when powdered, has been poached to near-extinction levels to satisfy man's desire for improved sexual performance.

Man also kills many other wild animals, but he causes most of the deaths indirectly. Animal societies are decimated by his toxic chemicals, his use of natural habitats for farming and his destruction of balanced ecosystems in the building of roads, dwellings and factories. The animals eaten by man today are bred under controlled conditions, and with an ever-increasing use of technology, to provide food. Predation has become a business, killing is one step removed from the consumer.

THE WORLD'S HUNTING PEOPLES

1 Eskimos
2 Ojibwa
3 North Algonkians
4 Micmac
5 Penobscot
6 Maidu
7 Seri
8 Alakaluf
9 Yaghan
10 Pygmies
11 Akoa
12 Bushmen
13 Birhors
14 Chenchu
15 Kadar
16 Andamanese
17 Semang
18 Yümbri
19 Aborigines
20 Ainu

TOP PREDATORS: UNCHALLENGED KILLERS

The African Lion

The sun, source of all energy, begins all food chains by its sustenance of plant life. At the far end of the feeding relationships begun there are the top predators, the animals that are rarely, if ever, preyed upon (except by man). Chains of feeding relationships vary in complexity but, as far as predation is concerned, they all end with the top predators.

Top predators tend to be large members of their respective groups. This arises from the fundamental fact, known as "the principle of food size", that predators usually have to be bigger than their prey in order to subdue them successfully. Therefore in a food chain moving from plant-eaters up through a succession of flesh-eaters eating other flesh-eaters, there is a series of animals of increasing size. Only by possessing particularly effective attacking weapons or by hunting in groups are predators able to reverse this size principle and kill animals bigger than themselves.

Because they are at the tip of a pyramid of animal populations built according to the rule that prey animals must always outnumber their predator, top predators are inevitably scarce in their environments. There are always more wildebeest than lions in the plains of East Africa and vastly more lemmings than snowy owls in the icy tundra of Scandinavia. Many top predators are on the danger list for extinction because of this unavoidable biological rule, as their small numbers make them vulnerable to changes forced on their habitats by man. Lions, for example, were once found in Southern Europe, eastwards through India and throughout the whole African continent. Now, the only wild lions outside Africa are in the Gir Forest in India and, even in their African heartlands, lions have been gradually pushed back from the north and south of the country until they are now only seen in their previous densities in the great plains of East Africa.

Most top predators have catholic tastes. As they are larger than most prey animals their range of attainable food is vast and many take full advantage of this. A lion will eat a zebra, a beetle and most of the animals between the two.

Strongly territorial, the top predator needs to have an area for its own predation which is large enough to sustain it. The individual or group will know this territory intimately and so hunt more efficiently in it. Hunting territories tend to be neatly organized alongside one another with little overlap. The ecologist Maurice Hornocker has shown how mountain lions in the wild upland country of Idaho mark their individual hunting areas with scrapes and excrement. Where they overlap the lions seem to take great care not to interfere with each other's kills.

To achieve the position of a top predator, an animal must be armed with lethal weapons such as specially adapted teeth and claws. It must also be so powerful and well protected itself that it is unlikely to be attacked. The shark's tough skin, the eagle's keen sight, the lion's sheer strength are structural features which make an animal a top predator.

Modern biologists no longer call him "King of the beasts" but the lion, *Panthera leo*, is still the unchallenged top predator of the plains and scrubland of Africa. The maned males may be 9 feet (2.75 metres) long, 3.5 feet (1 metre) high at the shoulder and weigh up to 550 pounds (250 kilograms). Equally powerful are the smaller, sleek-headed lionesses, about 8 feet (2.5 metres) long. With tigers, these cats are the biggest of the true land-hunting carnivores.

An adult lion needs to kill the equivalent of one wildebeest a week—the average wildebeest weighs 238 pounds (108 kilograms)—to survive. All parts of the animal are eaten except for any stomach contents, scraps of bone, and teeth. Lions often feed on the intestines first, sometimes even eating these alone and leaving the rest for scavengers. This seeming preference may be to supply fats and vitamins necessary to the lion's diet.

Powerfully-muscled hind limbs provide the thrust for the final spurt of an attacking lion. A lion can run for short bursts at about 40 mph, make standing jumps up to 12 ft (3.5 m) high and horizontal leaps of 40 ft (12 m).

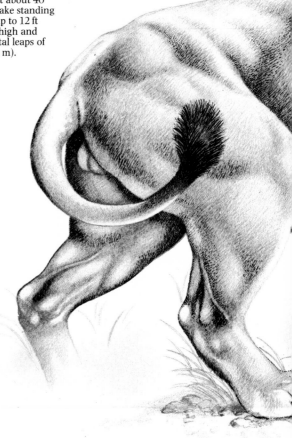

This male lion has brought down an impala. His next move will be to destroy the animal with a few powerful bites before devouring it. Once caught, many animals struggle surprisingly little and even seem to be in a state of shock, lying quietly before their mighty attacker with no means of defence.

THE SKULL OF A LION

Lions' teeth incapacitate a prey animal with a few destructive bites. The prime offensive weapons are the four pointed canine teeth which are capable of inflicting terrible penetrating wounds. Between the canines are small incisor teeth to chop mouthfuls of flesh from a kill. The skull itself is a massive structure with powerful jaw-closing muscles. Male and female skulls are similar but the outline of the male lion's skull is obscured and increased by the huge mane.

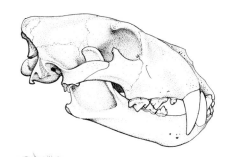
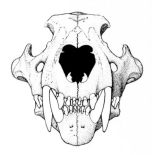

Tiny sharp backward-pointing hooks cover most of the upper surface of the lion's tongue. These form an efficient rasp for scraping meat fragments from a carcass or for grooming fur.

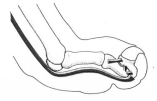

A vicious clawed weapon converts to a tough padded paw for running on because each outer toe bone and claw can fold right back into a fleshy sheath.

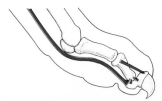

A strong elastic ligament holds the claws when retracted. To extend the claws, flexor muscles straighten the outer toe bones and thus protrude the claws.

The Lioness

Of the forty species of cat around the world only the lion lives and hunts in groups. Tigers, leopards and jaguars are solitary hunters of forest regions, but the lion has a significant social life on the dry plains that are its home. A pride may have only three or four members, or as many as one pride in East Africa observed to have two males, ten lionesses and 24 younger animals. Biologists believe that the group activities of the lion have evolved in response to two types of pressure, both linked with their hunting behaviour.

The first pressure is connected with the frailty of their young. Lion cubs, born blind and helpless, and unable to hunt for themselves for a year, are extremely vulnerable to predation by jackals and hyenas. In a pride, however, some lionesses can guard the cubs while others are off on the hunts.

The second pressure which has pushed lions into living in groups is the need to kill some animals bigger than themselves to give them a sufficient range of prey. An adult buffalo, for instance, may be twice the weight of a single lion, a giraffe three times as heavy, so the cooperative tactics of a group of lions are essential to hunt and kill these enormous creatures. In most prides group hunting is the responsibility of lionesses.

Although lions will kill in daylight given the opportunity, the hunt usually takes place under cover of darkness. Keen sight and hearing are the most important senses in the detection and ambush of prey. One or more individuals make a hidden approach which may last for several hours until they are close enough to make a short rush at their quarry. They seize the prey with one clawed paw on its back and the other on its flank, or grab the animal's rump with both forepaws to pull it over. Once the prey is down, a lioness lunges at the throat, bites into it and holds on until the animal dies from suffocation. Killing methods vary but an attack on the vulnerable head and throat area quickly incapacitates most prey.

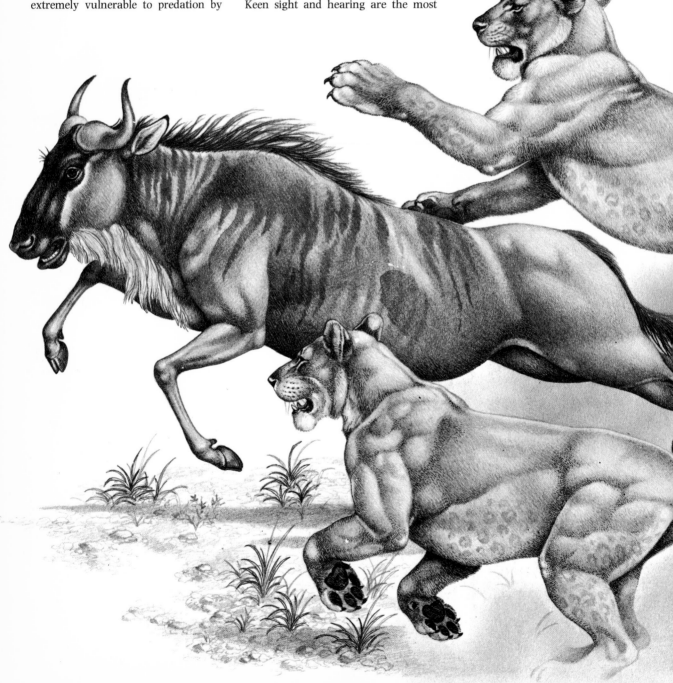

Markings on the back of a lioness's ears are "flashed" as the ears are lifted and used as a signal in the coordination of group activity.

SIGHT CAPACITY

Excellent vision is crucial for the detection of prey and for the accuracy of the final killing rush. Compared with the herbivorous rabbit or omnivorous man, the lion has a greater zone of stereoscopic vision.

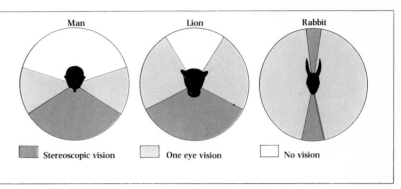

Man   Lion   Rabbit

■ Stereoscopic vision   ▨ One eye vision   □ No vision

Each sensitive white whisker on a lion's muzzle extends from a dark spot of fur. No two lions in an area will have the same spot patterns.

The modern lion, with fewer teeth than its ancestors, has developed a relatively short muzzle. This structure and the lion's large forward-pointing eyes allow the fields of vision of the two eyes to overlap, giving a wide sector of stereo-scopic binocular vision.

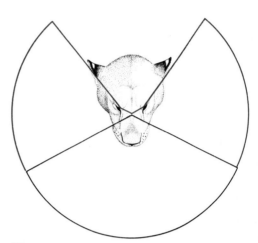

The animals killed by lions vary from area to area and with time even in the same place. The prey taken certainly depends on what is available, but lions do not simply kill in direct proportion to the numbers of different prey species. To demonstrate the quantity of prey a lion needs to survive the diagram shows the number of animals killed by one pride of ten lions in an average year.

■ 70 Wildebeest
▨ 50 Zebra
▨ 160 Thomson's gazelle
■ 8 Buffalo

■ 18 Topi
▨ 7 Warthog
□ 4 Grant's gazelle
▨ 11 Others (birds, reedbuck etc.)

A GROUP KILL

The lions sight a herd of prey. Several lionesses hide down-wind. Others circle upwind and stampede the herd towards the crouched lionesses waiting to kill.

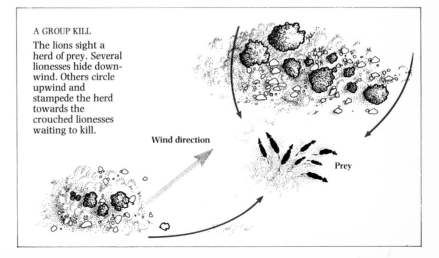

Wind direction

Prey

The lion is not a fast runner over long distances—a fact which makes the speed of many of the hoofed animals it kills their most important defence. The wildebeest, *Connochaetes taurinus*, under attack by two lionesses, is a key grazing animal in the short grasslands and open savannah of Africa and a common lion prey.

The Golden Eagle

Universally regarded as the king of the bird world, the eagle is the epitome of a top bird predator. Eagles are, without exception, carnivorous and rarely have any enemies other than man. Unfortunately, their slow breeding rate and their vulnerability to toxic chemicals mean that many species are declining in numbers. The American bald eagle *Haliaeetus leucocephalus* is one example.

Eagles are part of the hawk family, Accipitridae, with buzzards, kites, harriers and true hawks. These daytime hunting birds of prey—the other group is the falcon family—are known as raptors. Species occur throughout the world, feeding on reptiles, mammals and birds. The Australian wedge-tailed eagle, *Aquila audax*, hunts rabbits but

because it occasionally kills lambs it has been so ruthlessly shot and poisoned that its population is drastically reduced. The Verreaux's eagle, *Aquila verreauxi*, in South and East Africa, preys on birds as large as a guinea fowl, but its main food is the rock hyrax, a bizarre, rabbit-sized, distant relative of the elephant.

Most is known about the golden eagle, *Aquila chrysaetos*, native of North America, Scandinavia, North Africa, Europe and parts of Asia. It inhabits desolate mountainous areas and builds its nest, or eyrie, from sticks and heather on trees or cliff ledges. Nearly 3 feet (90 centimetres) long, it has a wing-span of up to 7 feet (2 metres); females are slightly larger than males but have

similar plumage. The eagle's exceptionally keen eyes can perceive details in a landscape eight times smaller than man can manage.

The hunting technique of the eagle depends on its ability to soar for long periods searching out its food. Prey is spotted from a great distance and then the bird makes a lightning dive to the ground to seize and kill the animal with its crushing talons. Using its huge hooked beak, the eagle rips its prey apart for eating. Hares, rabbits, grouse, geese and other birds are the eagle's usual food, as well as the occasional lamb or young deer.

Superbly designed for soaring and gliding flight, the eagle stays aloft for long periods with the minimum of energy expenditure, making long, shallow, downward glides, then spiralling upwards with thermal updraughts. The final rapid attack dive is initiated from flight or from a high vantage point on a mountain or tree.

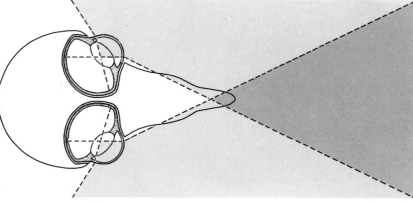

The true size of the eagle's eyes, much larger than their outward appearance suggests, is revealed by the vast orbit holes in the skull. The eyes point both forwards and sideways. At the front, the eagle has a wide sector of binocular stereoscopic vision, vital for the long-distance sighting of prey.

The upper bill of the eagle's beak is hooked and pointed at the tip for tearing at the flesh of its prey. This hook 4 in (10 cm) long fits over the flattened spoon-like ending of the lower bill.

The eye of a prey bird such as a duck is much smaller than that of an eagle and cannot provide the same detailed picture.
A highly curved cornea allows plenty of light to enter the eagle's eye and sensitive zones of the retina (fovea) have special light-sensitive cells which give the most precise visual perception.

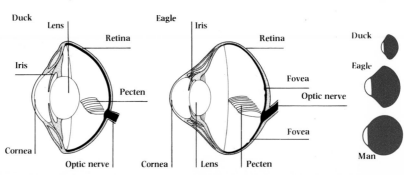

Duck

Lens

Retina

Iris

Pecten

Cornea

Optic nerve

Eagle

Iris

Retina

Fovea

Optic nerve

Fovea

Cornea

Lens

Pecten

Duck

Eagle

Man

The forelimb (wing) of a bird is an extreme modification of the normal vertebrate bone structure of the arm and hand which forms a basis for the attachment of wing feathers.

Fused carpals and metacarpals

Digit II

Alula

Digit III

Digit IV

Alula

Humerus

The eagle's wings have to support large loads when the bird carries its prey back to the nest. The structure of the feathers prevents "stalling", or sudden breakdown in the lifting power of the wings due to a loss of smooth airflow around them.

The main wing feathers, primaries, are tapered at the tip unlike the lower wing feathers. When fanned out they form a stack of staggered winglets, each acting as a slot device for the feather behind it.

Huge talons have the killing power an eagle needs to subdue a large prey animal. All four toes have long curved claws.

The bastard wing, or alula, forms a leading-edge slot which squeezes the airflow back on to the top surface of the wing instead of allowing it to break away as vortex eddies.

The Great White Shark

The biggest and most terrible of the predatory fish are the large sharks. The great white shark, *Carcharodon rondeleti*, has the most bloodstained reputation of all, for it is the maneater. A huge and powerful underwater killer, the great white shark can be 40 feet (12 metres) long from the tip of its snout to the end of its 10-foot (3-metre) high tail.

Found in all tropical oceans, this shark occasionally migrates into more temperate seas. Almost anything that swims or falls into the sea may be its prey: human heads, full-grown dogs, empty cans and mutton bones have been found in the stomachs of great white sharks. But large fish, turtles and

sea lions are the shark's more usual diet.

The shark's prime weapon is its set of jagged-edged, triangular teeth which hold and cut through flesh. Once the prey is in its jaws, the shark makes powerful rolling movements by waving its pectoral fins. Water resistance and inertia hold the prey steady while the cutting teeth slice through the prey like a chainsaw as the shark moves. The teeth in a pair of great white shark jaws in the Natural History Museum in London are 3 inches (7.5 centimetres) long; their owner was 36 feet (11 metres) long.

The sleek, streamlined body of the great white shark, as that of other

sharks, cuts through water at high speed (a blue shark was timed at 43 miles per hour). The strong supple skeleton is made from bendy cartilage rather than the bone in other fish. There are five gill slits at the rear of the head which are visible externally.

The shark's skin has the consistency of chain-mail armour. Hundreds of thousands of tiny flattened enamel spines cover the skin, each anchored by a bony root. These spines, called denticles, provide tough outer covering and protect the skin, and therefore the shark, from attack by all but the biggest predators, such as whales, or the weapons of man.

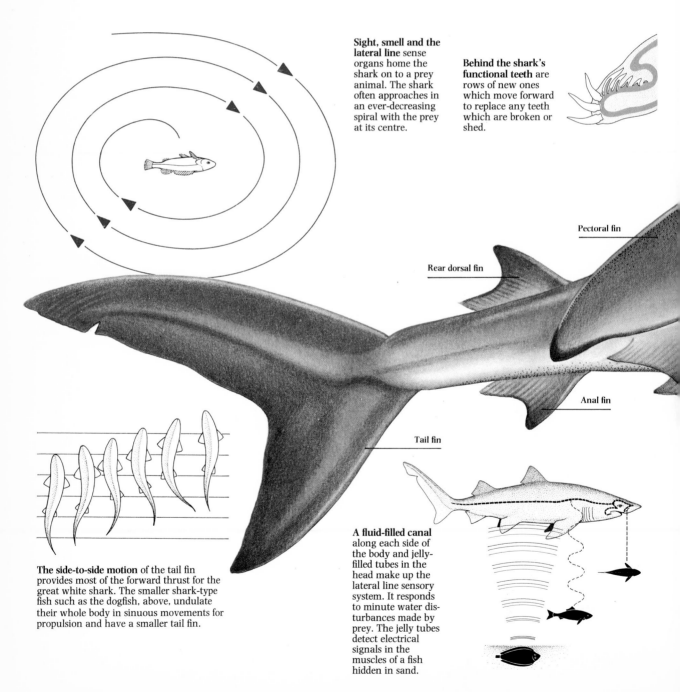

Sight, smell and the lateral line sense organs home the shark on to a prey animal. The shark often approaches in an ever-decreasing spiral with the prey at its centre.

Behind the shark's functional teeth are rows of new ones which move forward to replace any teeth which are broken or shed.

Rear dorsal fin

Pectoral fin

Anal fin

Tail fin

The side-to-side motion of the tail fin provides most of the forward thrust for the great white shark. The smaller shark-type fish such as the dogfish, above, undulate their whole body in sinuous movements for propulsion and have a smaller tail fin.

A fluid-filled canal along each side of the body and jelly-filled tubes in the head make up the lateral line sensory system. It responds to minute water disturbances made by prey. The jelly tubes detect electrical signals in the muscles of a fish hidden in sand.

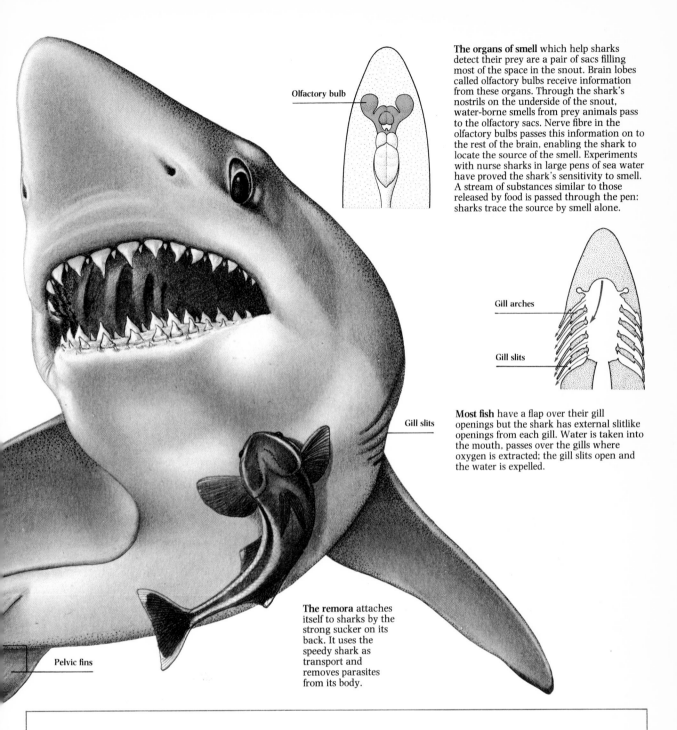

Olfactory bulb

The organs of smell which help sharks detect their prey are a pair of sacs filling most of the space in the snout. Brain lobes called olfactory bulbs receive information from these organs. Through the shark's nostrils on the underside of the snout, water-borne smells from prey animals pass to the olfactory sacs. Nerve fibre in the olfactory bulbs passes this information on to the rest of the brain, enabling the shark to locate the source of the smell. Experiments with nurse sharks in large pens of sea water have proved the shark's sensitivity to smell. A stream of substances similar to those released by food is passed through the pen: sharks trace the source by smell alone.

Gill arches

Gill slits

Most fish have a flap over their gill openings but the shark has external slitlike openings from each gill. Water is taken into the mouth, passes over the gills where oxygen is extracted; the gill slits open and the water is expelled.

Gill slits

The remora attaches itself to sharks by the strong sucker on its back. It uses the speedy shark as transport and removes parasites from its body.

Pelvic fins

THE SHARK'S JAW MECHANISM

The superbly streamlined shape of the shark, with jaws flush with the underside of the head, suggests a mouth incapable of opening wide enough to take big prey.

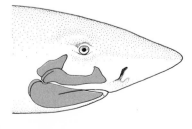

In fact most species of large sharks can drop their lower jaw until it is at right angles to the skull, making a deep, tooth-fringed scoop. The upper jaw cartilages

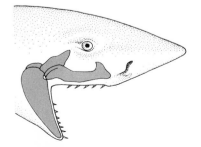

are tethered to the cranium by ligaments which ensure that the jaw is not wrenched too far open by a prey animal prepared to fight.

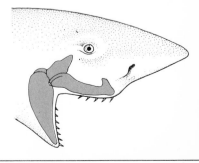

The Nile Crocodile

The crocodile's archaic appearance might suggest it is a modern version of the dinosaur and, in fact, its history does stretch back 150 million years. With alligators, caimans and gavials the crocodile forms an order of reptiles called the Crocodilia.

Only about 25 species still exist. Almost all of these are amphibious. freshwater predators in tropical areas of the world. The true crocodiles, such as the Nile crocodile, are the most widespread occurring in Central America, Africa, Asia, the East Indies and North Australia.

Alligators have shorter broader heads than crocodiles and can be distinguished by a lower tooth which penetrates into a hole in the upper jaw bone. Most are in Central and South America

but there is one species in China. Caimans, close relatives of alligators, also live in tropical Central and South America. These three short-snouted crocodiles kill a wide range of animals for food and frequently catch their prey on land. Gavials, found in the major rivers of India, are underwater hunters and prey on fish. Their long tapering jaws are ideal for seizing slippery fish.

The full-grown male Nile crocodile, *Crocodilus niloticus*, can be up to 16 feet (5 metres) long and is capable of overcoming animals as large as waterbuck and buffalo. The well-armoured body is almost completely covered with horny scales. Rows of thickened, pitted bony plates on the crocodile's back provide additional protection. The softer underbelly is shielded by bony out-

growths called abdominal ribs which develop in the abdominal wall.

The Nile crocodile drowns its prey by holding it under water and subtle physical modifications help it in this method of killing. Lids and flaps cover the eyes and ears under water and the mouth cavity can be closed off by a flap valve formed from the base of the tongue and the rear part of the palate. Thus the crocodile can breathe through the nostrils and palate airway even when the jaws are under water.

Once widely distributed in Africa from the Nile delta in the north to South Africa, the Nile crocodile is now reduced in numbers and range—because of the destruction of former habitats and the commercial exploitation of its skin for the manufacture of luxury goods.

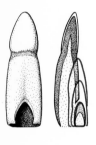

The pointed teeth are embedded in deep sockets in the crocodile's jaw bones. Throughout the crocodile's life, new teeth develop to replace worn ones. The new tooth starts in a cavity at the base of the tooth it is to replace and grows until it pushes out the old tooth.

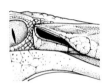

The ear drum is close to the skin's surface and is covered while under water by a flap of tough skin.

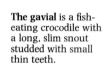

The gavial is a fish-eating crocodile with a long, slim snout studded with small thin teeth.

The alligator's canine-like teeth actually lodge into holes in the upper jaw bones when the jaws are shut.

The crocodile's large canine-like teeth on the lower jaw fit into notches between the upper teeth when the mouth is shut.

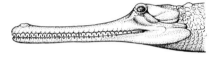

Nostril

Olfactory organ space

Hard palate

Tube to lungs

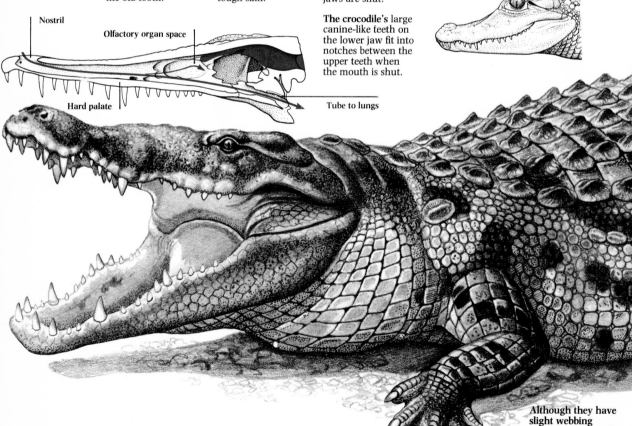

Although they have slight webbing between the toes, the crocodile's feet are used mainly when moving on land.

HOW A CROCODILE FEEDS

The crocodile must tear its food into pieces before eating because the structure of its head—the solid unyielding palate of bone and the fixed hinges linking the lower jaw to the skull—prevents it swallowing large animals whole. By holding on to its prey and spinning its own body round, the crocodile twists off limbs or convenient chunks of flesh.

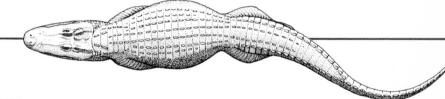

The crocodile deliberately swallows stones which act as stabilizers when swimming. The stomach of an adult crocodile contains up to 11 lb (5 kg) of stones.

Undulations of the crocodile's elongated, flattened tail provide a powerful swimming thrust. The limbs are kept close to the sides in water.

The stomach contents of crocodiles at different ages and stages of development were examined to discover how the diet changes through the crocodile's life. The diagram shows the frequency with which various types of prey were found.

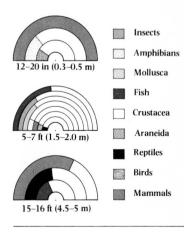

12–20 in (0.3–0.5 m)

5–7 ft (1.5–2.0 m)

15–16 ft (4.5–5 m)

Insects
Amphibians
Mollusca
Fish
Crustacea
Araneida
Reptiles
Birds
Mammals

The Nile crocodile, *Crocodilus niloticus,* is the biggest of the three types of crocodile found in Africa. A stealthy hunter, the crocodile attacks prey on land or in the water. It will lurk near the shore waiting for an animal to come and drink. With a sudden burst of speed, the crocodile seizes the prey by the nose or leg, drags it into the water and holds it until drowned. The prey taken changes radically as the crocodile develops. The young live on insects and spiders. As the crocodile grows, fish, crabs and snails are its main diet and the fully-grown adult feeds on large animals and birds.

CAPTURE BY CONCEALMENT: HIDDEN DANGER

Visual Deceits

In a world where vision is a crucial sense for most animals, it can be a huge advantage to be inconspicuous. Being hard to see can help hide predators from the animals they hunt or aid quarry to elude their pursuers. But being inconspicuous is useless if combined with ostentatious movement and the making of loud noises—truly effective camouflage is a mixture of both body structure and behaviour.

For predatory carnivores, good camouflage confers two sorts of advantage. It can make ambushing a tenable technique and, for predators that stalk and chase their prey, can allow the hunter to close in nearer to its intended victim before the prey starts to run. Even a few feet so gained can tip the balance of the chase in the predator's favour. And because most predators are themselves prey to yet larger predators, camouflage gives added protection against capture.

The patterns of predator camouflage are almost as varied as the animals themselves. For maximum gain, many predators use a combination of the basic camouflage methods. This is particularly true of predators whose camouflage must be as perfect as possible, and the end result often involves close mimicry of some part of their habitat or even of another, harmless animal which the prey will allow to approach.

Because many prey animals have colour vision, any predator will have difficulty in coming shoulder-to-shoulder with its prey if its colour contrasts with the landscape. The acquisition of body colours similar to those of their backgrounds is thus a common means of camouflage. Like the white snowy owls and polar bears of the Arctic, animals inhabiting environments of monotonous colour find refuge in being almost solidly that colour. When the habitat has a patchy colour distribution, camouflage involves changing hue. Many lizards, fish and cephalopods (for example, the cuttlefish) use this colour-matching ability to gain the predatory upper hand.

The strategy of counter-shading is a camouflage device widely used by predators. Clues about the three-dimensional shapes of objects come from the patterns of light and shade on their surfaces and solid objects lit to reveal these patterns stand out from their backgrounds. Normally, with light coming from above, the top of an animal is lighter than the shaded underside, but in counter-shaded predators such as the mackerel, light and dark colours are reversed, neatly offsetting the norm and making the animals appear flat and inconspicuous from every direction.

Projected shadows will also make a creature conspicuous and betray its position. Some predators have such sophisticated control of light, shade and shadows that their body shapes are adapted to eliminate virtually all shadows. Most, like the skates and rays, have flattened shapes, with any edges that could make shadows smoothed out by flaps on their sides.

The characteristic outline of a predator is often known and feared by its habitual prey. Predators like the leopard and tiger have bodies patterned with spots, blotches or stripes. This disruptive camouflage breaks up the instantly recognizable shape and leads the eyes of the prey away from the silhouette to the markings.

Shape and colour are the physical ingredients of successful camouflage. Changes in body shape are the most extreme moves in the camouflage game, for alteration in body architecture may well reduce overall efficiency. Such changes thus have a limited range, but flattened bodies, like those of the plaice, which remove shadows and prevent an animal standing out from its surroundings, are common. The gecko and carpet shark are among the many predators which fuse with their backgrounds by

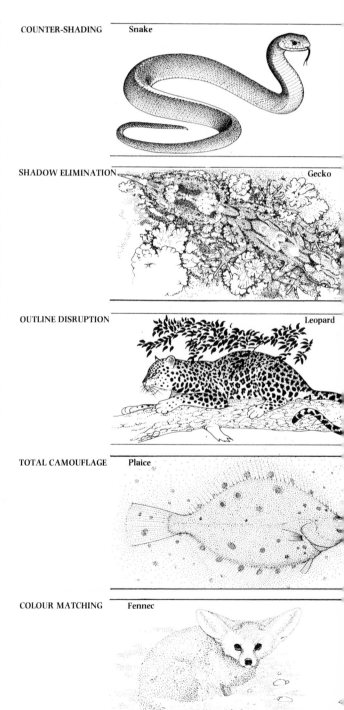

COUNTER-SHADING   Snake

SHADOW ELIMINATION   Gecko

OUTLINE DISRUPTION   Leopard

TOTAL CAMOUFLAGE   Plaice

COLOUR MATCHING   Fennec

virtue of flaps or appendages covering shadow-producing body areas. Usually only non-essential body parts such as skin and cuticle are used for these flaps.

Camouflage by colour and pattern imposes fewer constraints on an animal. The great variety of light, shade, colours and patterns is produced in several ways. Intrinsic colours and markings are the most common and are created by pigments in the skin, usually housed in special cells. Chameleons and cuttlefish are among the animals whose pigment cells can expand and contract to effect quick colour changes. Interference colours—glossy iridescent mixes—appear when light passes through or bounces off finely divided surface structures, or is reflected from a thin layer of surface material.

An extreme example of predatory camouflage is the sabre-toothed blenny which uses its close resemblance to the harmless cleaner wrasse to get close to unsuspecting fish. It then attacks and bites its prey.

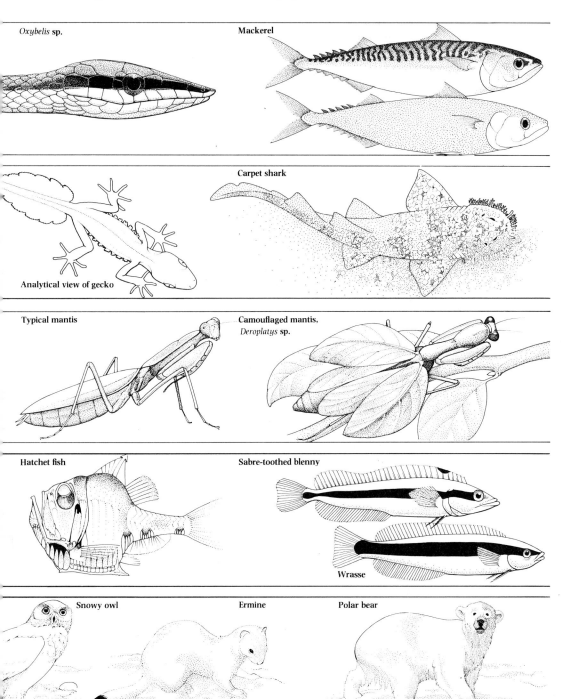

Oxybelis sp.

Mackerel

Analytical view of gecko

Carpet shark

Typical mantis

Camouflaged mantis, *Deroplatys* sp.

Hatchet fish

Sabre-toothed blenny

Wrasse

Snowy owl

Ermine

Polar bear

Most snakes are counter-shaded, which allows them to get close to their victim without being seen. In some, such as *Oxybelis*, the dark upper colour runs through the eye area, making it less obvious. Many predatory fish are counter-shaded.

The Malaysian flying gecko typifies the "flap" technique of shadow elimination. It has flaps on the sides of its body and tail so that it "disappears" when flattened against a tree trunk. The carpet shark, with its frilly edges, achieves the same trick.

The leopard, with its characteristic spots, is a predatory example of outline disruption, but prey such as zebras share this technique. Outlines can also be physically changed —some mantis species conceal their typical outline and resemble leaves.

Plaice are flattened, and can change colour. Hatchet fish are laterally flattened and have silvery scales set to correspond to background light. The sabre-toothed blenny uses its resemblance to the harmless cleaner wrasse to get close to prey.

Desert and arctic animals are good examples of colour matching camouflage and many of these predators are the same colour as their habitat. The fennec fox of the Sahara is a sandy yellow, the polar bear, snowy owl and ermine are white.

The Indian Tiger

"Every tiger has its kill." So runs the old Chinese proverb—and it is true. The most handsome and lordly of the big cats, the tiger is a solitary, territorial hunter, unlike its near relative, the lion, which pursues its prey in a coordinated pride. In other, physical, respects, the lion and the tiger, which belong to the same genus, *Panthera*, are extremely similar. It is said that when skinned the two animals are almost identical structurally.

While lions are the top predators of open country, tigers thrive where there is more cover. A protective surround of vegetation is crucial to the tiger's hunting strategy, for among trees and bushes the striped, black-brown markings on its yellowish coat provide excellent camouflage by breaking up the cat's silhouette. This is the tiger's greatest asset in the stalking phase of the hunt.

Tigers inhabit environments ranging from the snows of Siberia to the steamy tropical rain forests of the East Indies,

and have evolved into seven distinct races, each adapted to a particular habitat. These are found in Manchuria, the Caspian Sea area, China, India, Sumatra and Java. Although they may interbreed at the margins of these zones, the tigers maintain some differing physical characteristics in order to be successful predators in their own environment. Because of these varied living conditions, tigers feed on a large selection of prey. They have been known to kill animals as large as wild bull buffaloes, will take many species of deer, and antelope, as well as wild pig, monkeys and porcupines, but may also eat minute prey like locusts. Excellent swimmers, tigers will capture fish and turtles in marshy country.

Using stealth, cover and the inconspicuousness its camouflage affords, the tiger stalks very close to its chosen prey which, in dense vegetation, is located by a powerful sense of smell and acute hearing rather than by sight. Once

within attacking range, the big cat rushes at its quarry over a short distance. Usually it grasps at a shoulder with one clawed foot, then seizes the prey's throat from the underside. This manoeuvre can break the neck of small prey—larger ones probably die from suffocation as the windpipe is compressed or crushed. After a successful kill, the tiger almost invariably carries or drags its victim to a secluded spot in dense cover.

When game is in short supply, tigers may turn to easier pickings, but in doing so they come into immediate conflict with man. Hungry or diseased tigers kill domesticated cattle and the most notorious of them become man-eaters. Man in turn affects the tiger population by hunting or destroying its habitats for housing or agriculture. As a result, the tiger is now high on the list of endangered species. The race on the island of Bali is almost certainly extinct and other species are in danger.

Like a domestic cat stalking a bird, the tiger uses every trick of positioning its sinuous body to get close to its prey. By creeping straight towards the prey animal and crouching close to the ground, the tiger reduces the area displayed to it.

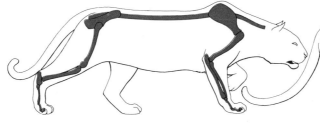

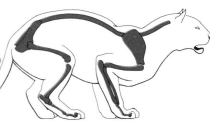

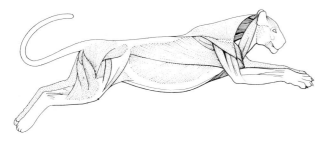

The final move in the tiger's attack on prey can be in the form of a rush or a leap depending on the type of terrain. This powerfully-muscled animal is on record as having made a single leap of 18 ft (5.6 m) from the ground to pull a man from a tree.

The prominent dark stripes on the tiger's coat must be the most familiar example of disruptive camouflage. Their function is to draw the prey animal's eye away from the tiger's body shape, which it would normally recognize and react to. The stripes enable the tiger to conceal itself in small clumps of vegetation and blend into the background.

The stripes never coincide with the real outline of the tiger's body but always run at right angles to it. An artificial tiger with stripes running with its outline would be hopelessly obvious and unable to camouflage itself successfully.

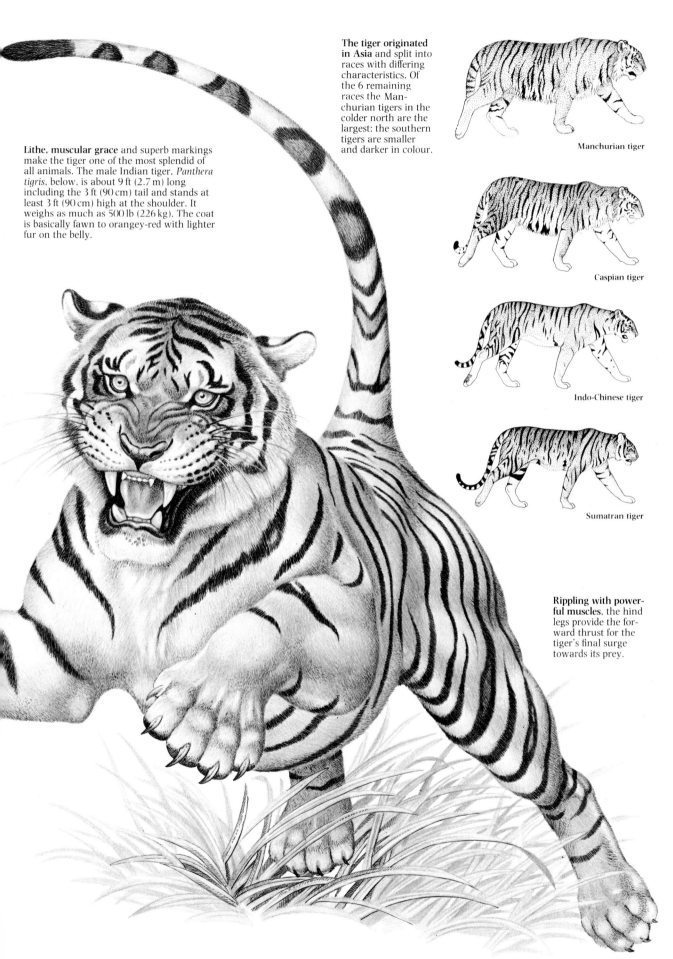

The tiger originated in Asia and split into races with differing characteristics. Of the 6 remaining races the Manchurian tigers in the colder north are the largest; the southern tigers are smaller and darker in colour.

Manchurian tiger

Caspian tiger

Indo-Chinese tiger

Sumatran tiger

Lithe, muscular grace and superb markings make the tiger one of the most splendid of all animals. The male Indian tiger, *Panthera tigris*, below, is about 9 ft (2.7 m) long including the 3 ft (90 cm) tail and stands at least 3 ft (90 cm) high at the shoulder. It weighs as much as 500 lb (226 kg). The coat is basically fawn to orangey-red with lighter fur on the belly.

Rippling with powerful muscles, the hind legs provide the forward thrust for the tiger's final surge towards its prey.

CAPTURE BY CONCEALMENT
The Praying Mantis

The praying mantids are a group of large insects that do not live up to their devout-sounding name. Each one of the 1600 known species is a rapacious predator. The name refers to the characteristic attitude adopted by these animals when they lie in wait for their prey, usually smaller insects. Their forelegs, deadly killing organs, are folded in demurely under the head so as to resemble the posture of someone at their prayers.

As soon as a fly appears within the reach of these front raptorial limbs, the praying mantis turns its head and strikes. The attack is lightning fast; the front legs reach out together and rarely miss their target. In extending forward, the legs reveal sharp-pointed spines on the rear surface of two adjacent limb segments. As these segments flex together again, the prey is trapped between the rows of spines. The victim is then pulled back towards the mantis's body as the legs are brought back to a folded configuration. Soon the prey is chewed and swallowed.

A complete attack sequence takes only 30 to 50 milliseconds. The prey has little chance; in laboratory experiments about 85 out of 100 flies that came near the front end of a hungry mantis were captured and eaten.

The strike is over so quickly that no correction can be made once the mantis has committed its front limbs to an attack. As a result, success depends on pinpoint accuracy of aim before the final lunge; this is achieved by two separate systems—the eyes, and patches of sensory hairs on the insect's "shoulders".

The massive, multi-faceted, compound eyes are fixed immovably in front of the head, pointing forwards and sideways. The mantis can rotate its head to point exactly at any fly that lands within range. The insect's bearing in relation to the body is therefore computed by the degree of head turning. The tufts of sensory shoulder hairs confirm this head position before the legs are fired toward the prey.

Observations of newly hatched mantids have shown that they feed themselves in the same way as adults do. They do not have to learn by trial and error the trick of striking in the right direction at the right instant. In these insects, the art of catching flies is inborn or instinctive.

Potential prey victims will obviously try to avoid such an efficient predator. To make it difficult for intended prey to notice them before those fatal legs strike home, mantids have a bewildering range of techniques for camouflaging themselves in their surroundings.

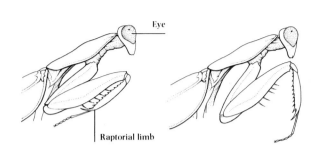

SEQUENCE OF A KILL
In the praying attitude the killing, or raptorial, limbs are bent back in a tight "N" shape. During the strike the "N" is opened out so that the limb is in an almost straight line extending towards the prey. The limb then folds back, catching the prey between spines on the femur and tibia.

Eye

Raptorial limb

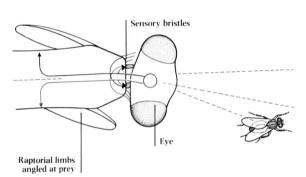

Sensory bristles

Eye

Raptorial limbs angled at prey

A superb aim is vital to the mantis's attack. First the head swivels like a tank turret to point at the target. This movement is used by the brain to compute the angle of the strike. Sensory hairs on the insect's shoulders provide a further check: as the head twists it distorts these hairs and from the pattern of distortion, the brain confirms the strike angle.

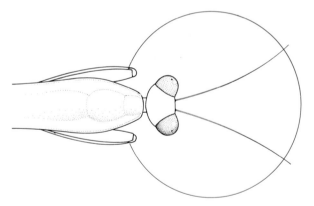

At the moment of the attacking strike, the mantis is still, having already stalked its prey with infinite care or waited in ambush. The mantis has a defined strike zone— the area in front of the raptorial limbs within which prey can be captured. The strike zone for the species *Hierodula crassa*, right, is 1 in (3 cm) across.

Remaining unseen until ready for the attack strike is an important part of the hunting technique of the mantis. Many species mimic leaves or twigs, like *Empusa fasciata*, below, and wait, perfectly concealed among the vegetation.

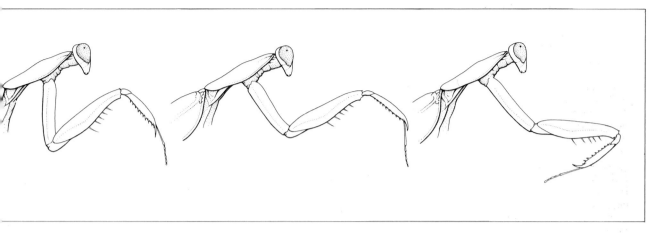

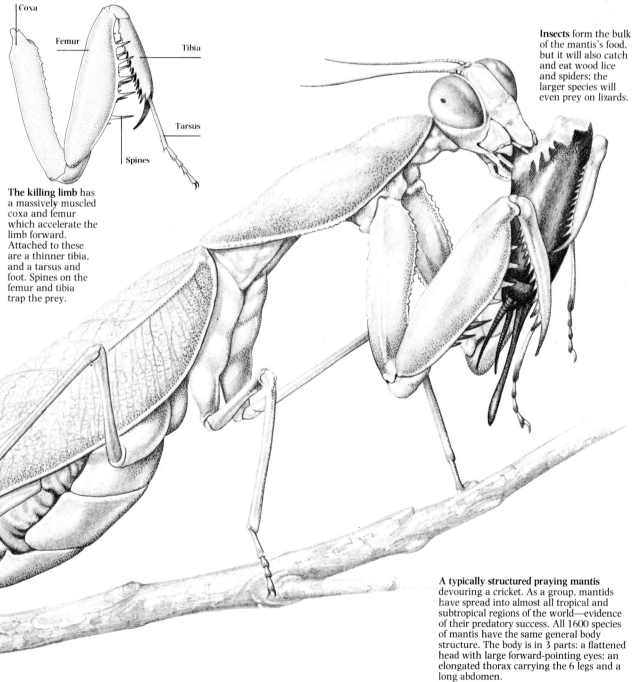

Coxa

Femur

Tibia

Tarsus

Spines

The killing limb has a massively muscled coxa and femur which accelerate the limb forward. Attached to these are a thinner tibia, and a tarsus and foot. Spines on the femur and tibia trap the prey.

Insects form the bulk of the mantis's food, but it will also catch and eat wood lice and spiders; the larger species will even prey on lizards.

A typically structured praying mantis devouring a cricket. As a group, mantids have spread into almost all tropical and subtropical regions of the world—evidence of their predatory success. All 1600 species of mantis have the same general body structure. The body is in 3 parts: a flattened head with large forward-pointing eyes; an elongated thorax carrying the 6 legs and a long abdomen.

The Chameleon

A tongue like a mobile fly-paper which catches and immobilizes prey in an instant—this is the chameleon's weapon. The chameleon is a highly modified member of the Lizard family, a group of reptiles which has adapted to a remarkably wide range of life styles. Different lizard species exist in deserts, burrow deeply underground, or live aquatic lives. Some lizards are nighttime insect feeders with a climbing ability so effective that they can scale a pane of glass; others have mastered gliding flight. No set of lizard specializations, however, is so extraordinary as that of the chameleon.

These reptiles have become perfectly adapted for a life of daytime predation in the trees and vegetation of tropical and sub-tropical forests. The feet and tail are so specialized for gripping branches that

the chameleon's movement on land is ungainly and not often attempted.

The chameleon's proverbial camouflage ability—coloured skin cells can change both its colour and pattern—is quite unequalled among lizards and serves two purposes. First, the chameleon's similarity to its background helps make its incredibly slow, stalking approach of prey animals like insects unnoticed. Second, other predators such as snakes attack chameleons, and remaining hidden is the best defence when an attacker is near. If actually attacked many species do have additional defensive responses. Some can blow their body up to a globular shape by an extreme expansion of the lungs and, with the almost instantaneous blackening of skin, this may well frighten off potential predators.

The chameleon's most bizarre adaptation is its most useful—the tongue which shoots out of its mouth to trap food. The tip of the tongue, which actually hits the prey, is an indented pad of sticky gland cells set in the tongue's surface. The sequence of extending the tongue, trapping prey and returning it to the mouth is all over in less than a twentieth of a second. The process has been the subject of much scientific controversy but modern ideas suggest that the tongue is fired along a tapering piece of bone by the rapid contraction of the tongue muscles.

About 85 species of chameleon are known to exist; about three-quarters of these live on the island of Madagascar. Among them is the world's largest chameleon, *Chameleo oustaleti*, which can be 2 feet (60 centimetres) long.

The initial precise aiming of its tongue strike depends on the chameleon's huge eyes. In the eyeball itself the cornea is small and the lens relatively flat, projecting a large image on the retina which is well separated from the lens. This means that the chameleon has high-fidelity vision over the small area at which the eye is pointing.

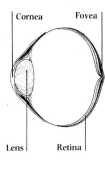

Cornea   Fovea

Lens   Retina

The eyes can work independently and point in quite different directions to cover most of the possible view on one side of the body. The cornea is located at the tip of a protruding cone of skin and, within this, muscles enable the eye to point almost 180 degrees forwards and backwards, almost directly downwards and, to an extent, upwards. Only a fly perched on the top of a chameleon's head could escape from these all-seeing eyes.

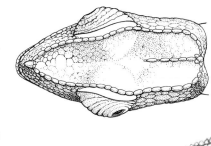

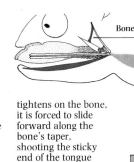

Insects, such as this member of the locust family (Orthoptera), are a common item in the chameleon's diet. Larger species, however, eat other lizards, birds and their eggs and even small rodents.

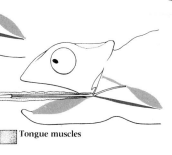

The lightning-fast tongue is operated by a complex battery of muscles. The retracted tongue is arranged like a telescoped muscular sleeve around a long, thin, tapering bone in the floor of the mouth. To fire the tongue, this bone is rocked forward by muscles; the muscle in the tongue contracts and, as it tightens on the bone, it is forced to slide forward along the bone's taper, shooting the sticky end of the tongue violently forward.

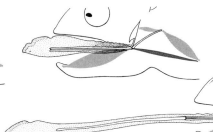

Bone

▢ Contracting muscles   ▩ Relaxed muscles   ▢ Tongue muscles

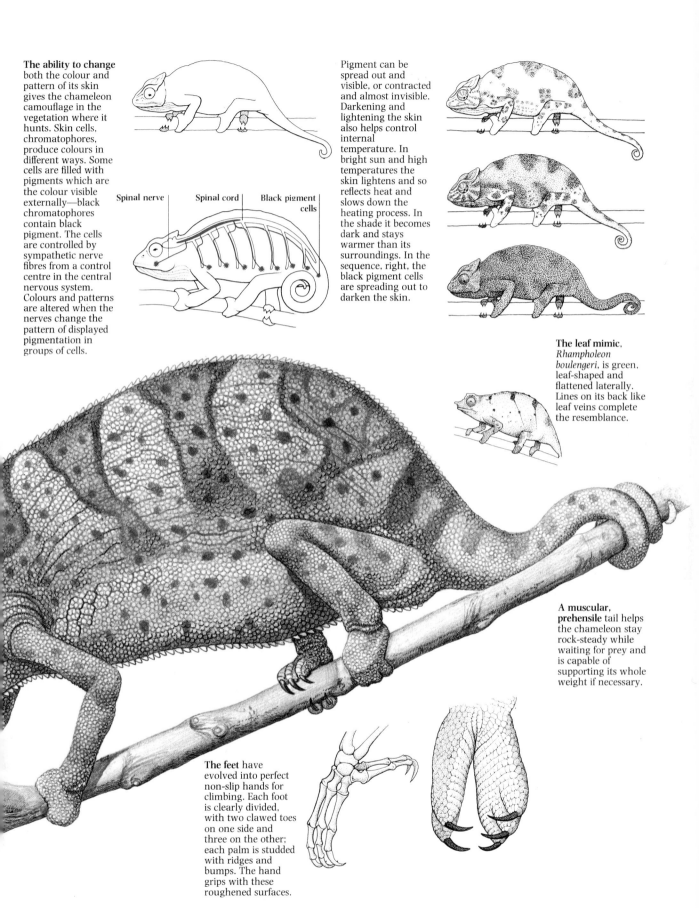

The ability to change both the colour and pattern of its skin gives the chameleon camouflage in the vegetation where it hunts. Skin cells, chromatophores, produce colours in different ways. Some cells are filled with pigments which are the colour visible externally—black chromatophores contain black pigment. The cells are controlled by sympathetic nerve fibres from a control centre in the central nervous system. Colours and patterns are altered when the nerves change the pattern of displayed pigmentation in groups of cells.

Spinal nerve   Spinal cord   Black pigment cells

Pigment can be spread out and visible, or contracted and almost invisible. Darkening and lightening the skin also helps control internal temperature. In bright sun and high temperatures the skin lightens and so reflects heat and slows down the heating process. In the shade it becomes dark and stays warmer than its surroundings. In the sequence, right, the black pigment cells are spreading out to darken the skin.

The leaf mimic, *Rhampholeon boulengeri,* is green, leaf-shaped and flattened laterally. Lines on its back like leaf veins complete the resemblance.

A muscular, prehensile tail helps the chameleon stay rock-steady while waiting for prey and is capable of supporting its whole weight if necessary.

The feet have evolved into perfect non-slip hands for climbing. Each foot is clearly divided, with two clawed toes on one side and three on the other; each palm is studded with ridges and bumps. The hand grips with these roughened surfaces.

With all four grasping feet and the curled tail maintaining an absolutely solid firing platform, a common chameleon, *Chameleo chameleo,* unerringly traps an insect. The left eye has been pivoted forward to aim the strike. The speed of the tongue strike (about 40 milliseconds) is such that the insect does not have time to make its rapid jumping response. The insect has been caught at the full extent of the projected tongue—over half the body length of the chameleon. The common chameleon is about 10 inches (25 centimetres) long including the tail and is found in parts of southern Europe, the Middle East and North Africa.

CAPTURE BY TRAPS: PREDATORS IN WAITING

Man is not alone in his use of a trap to capture prey. Many animals set traps and then wait passively for their quarry to ensnare themselves or set off the traps' killing mechanisms.

Large numbers of trappers are found among sessile animal species, those which have no independent means of locomotion and usually remain firmly attached in one place. Several invertebrates such as the corals and their close cousins the sea anemones and hydroids (marine-encrusting colonies of animals such as *Hydra*) fit into this category. Although the rock-like skeleton of the coral is static, it houses thousands of tiny sea-anemone-like individuals living a colonial cooperative life—each individual is a trapper. Small planktonic animals, mainly crustaceans and fish larvae, are carried by sea currents on to the coral's tentacles, which are equipped with stinging cells, where they stick and are paralyzed.

The corals and their free-floating relatives the jellyfish and the Portuguese man-of-war all use traps that are parts of themselves. Other animals construct traps out of inanimate materials. Among these are insects called ant-lions, which dig conical pits in the sandy soils of deserts and other dry regions. The pit has unstable sides so that when a small insect enters the pit, sliding sand pushes it to the bottom where the ant-lion with its fang-like mouthparts soon disposes of it. Yet another group of animals exists that make traps from their own secretions. Spiders are probably the best-known examples with their bewildering range of webs.

Lures, utilized less often in the animal kingdom than traps, are organs, structures and behaviour patterns used by predators to attract prey to them. The alligator snapping turtle of North America, with a filamentous extension on its tongue, lures unsuspecting fish into its mouth. Glowworm larvae of the New Zealand fungus gnat (*Arachnocampa*) live in silken tubes in the roofs of caves. From these tubes, they dangle "fishing lines" covered with sticky droplets that glow with a blue-green light. Flying insects, attracted to the glowing lines, find themselves stuck on to a droplet. The glowworm immediately pulls up the line and devours its prey.

The Portuguese man-of-war

To many people, any floating sea animal with stinging tentacles is a jellyfish. There is, however, one group of animals that fits this description but is only distantly related to the jellyfish (Scyphozoa)—the siphonophores.

The Portuguese man-of-war, *Physalia*, common throughout the tropical Atlantic, is the best-known siphonophore. Although each one looks like a single animal, it is in fact a super-organism, or colony, made up of hundreds of individual animals.

The largest individual is the float (pneumatophore) which is about 8–10 inches (20–25 centimetres) long. It is a bag of secreted gas with a grooved sail-like vane protruding from it, and its function is to keep the colony afloat. Beneath it, dangling under the water, are the colony's linked individuals (polyps) of which there are three specialized types.

First, there are feeding polyps (gastrozooids), which are small digestive bags, each with its own mouth. Second, there are tentacle polyps (dactylozooids), which are elongated threads that may hang up to 20 feet (6 metres) below the float and are studded with stinging cells. Last, there are reproductive individuals (gonozooids).

As the Portuguese man-of-war moves in mid-ocean currents, it fishes with its tentacles, each of which is capable of harpooning and paralyzing fish, crustaceans and other marine animals with the stinging cells. The tentacles contract and haul the prey up to the feeding polyps, which engulf and digest it. Nutrients are then passed to all the individuals in the superorganism from the feeding polyps. It might be said, in fact, that the Portuguese man-of-war practises "predation by committee".

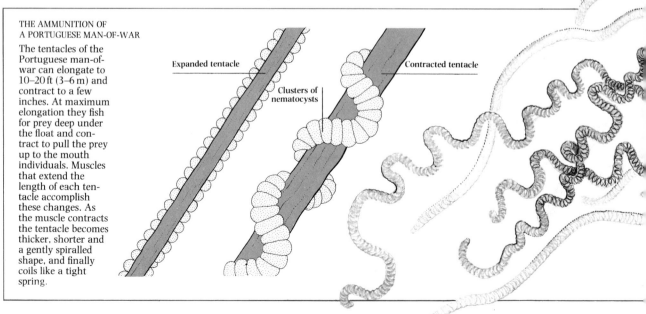

THE AMMUNITION OF A PORTUGUESE MAN-OF-WAR

The tentacles of the Portuguese man-of-war can elongate to 10–20 ft (3–6 m) and contract to a few inches. At maximum elongation they fish for prey deep under the float and contract to pull the prey up to the mouth individuals. Muscles that extend the length of each tentacle accomplish these changes. As the muscle contracts the tentacle becomes thicker, shorter and a gently spiralled shape, and finally coils like a tight spring.

Expanded tentacle

Clusters of nematocysts

Contracted tentacle

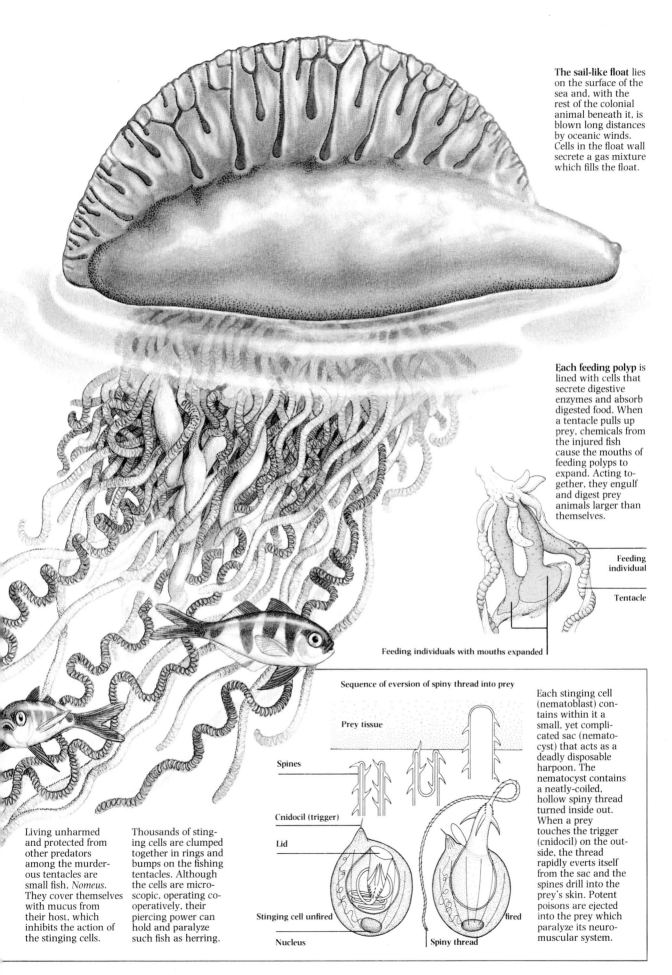

The sail-like float lies on the surface of the sea and, with the rest of the colonial animal beneath it, is blown long distances by oceanic winds. Cells in the float wall secrete a gas mixture which fills the float.

Each feeding polyp is lined with cells that secrete digestive enzymes and absorb digested food. When a tentacle pulls up prey, chemicals from the injured fish cause the mouths of feeding polyps to expand. Acting together, they engulf and digest prey animals larger than themselves.

Feeding individual

Tentacle

Feeding individuals with mouths expanded

Sequence of eversion of spiny thread into prey

Prey tissue

Spines

Cnidocil (trigger)

Lid

Stinging cell unfired

Nucleus

fired

Spiny thread

Living unharmed and protected from other predators among the murderous tentacles are small fish, *Nomeus*. They cover themselves with mucus from their host, which inhibits the action of the stinging cells.

Thousands of stinging cells are clumped together in rings and bumps on the fishing tentacles. Although the cells are microscopic, operating cooperatively, their piercing power can hold and paralyze such fish as herring.

Each stinging cell (nematoblast) contains within it a small, yet complicated sac (nematocyst) that acts as a deadly disposable harpoon. The nematocyst contains a neatly-coiled, hollow spiny thread turned inside out. When a prey touches the trigger (cnidocil) on the outside, the thread rapidly everts itself from the sac and the spines drill into the prey's skin. Potent poisons are ejected into the prey which paralyze its neuromuscular system.

The Trapdoor Spider

There are 20,000 species of spiders throughout the world, in a wide range of environments—under water, under ground, on the soil and among vegetation. Without exception, each is a fanged, carnivorous predator.

All spiders have the same basic structure and have adaptations for catching insects and other small animals. The body is divided into two parts: a forebody (prosoma) and larger hind body (opisthosoma) joined by a thin waist. Four pairs of walking legs extend from the forebody. In front of these are two extra pairs of short, leg-like appendages: the first pair, the fangs (chelicerae), are pointed, offensive weapons that can pierce the toughest

beetle cuticle and inject potent paralyzing poisons; the second pair (pedipalps) bear sense organs, which relay information about food, and bristles which strain food particles before they reach the mouth. Situated at the front of the forebody are two rows of four eyes, which are important, in those species that do not use webs, for finding prey.

The legs are flexed by internal muscles, but are extended by a hydraulic mechanism which powers fast running and jumping. To operate this mechanism the legs contain blood-filled spaces that are connected to other spaces in the forebody.

Spiders have an extraordinary specialization for which they are

probably best known—their ability to produce silk. This tough, fibrous mixture of proteins is used by spiders for every conceivable function: to gift-wrap sperm during copulation, to produce safety lines while jumping and to make thousands of differently designed traps and snares. Produced as a fluid in large glands that occupy much of the hind body, silk is then extruded as threads through complex nozzles called spinnerets near the rear end of the body. These threads can stretch up to 20 per cent of their initial length before breaking. In addition to this great elasticity, they are incredibly strong: a thread thinner than a strand of human hair can support a weight of 80 grams.

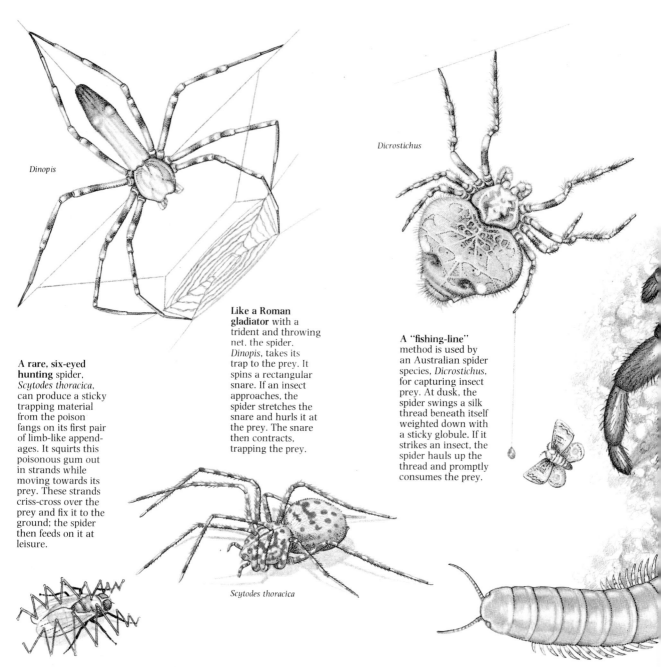

Dinopis

Dicrostichus

A rare, six-eyed hunting spider, *Scytodes thoracica*, can produce a sticky trapping material from the poison fangs on its first pair of limb-like appendages. It squirts this poisonous gum out in strands while moving towards its prey. These strands criss-cross over the prey and fix it to the ground; the spider then feeds on it at leisure.

Like a Roman gladiator with a trident and throwing net, the spider, *Dinopis*, takes its trap to the prey. It spins a rectangular snare. If an insect approaches, the spider stretches the snare and hurls it at the prey. The snare then contracts, trapping the prey.

A "fishing-line" method is used by an Australian spider species, *Dicrostichus*, for capturing insect prey. At dusk, the spider swings a silk thread beneath itself weighted down with a sticky globule. If it strikes an insect, the spider hauls up the thread and promptly consumes the prey.

Scytodes thoracica

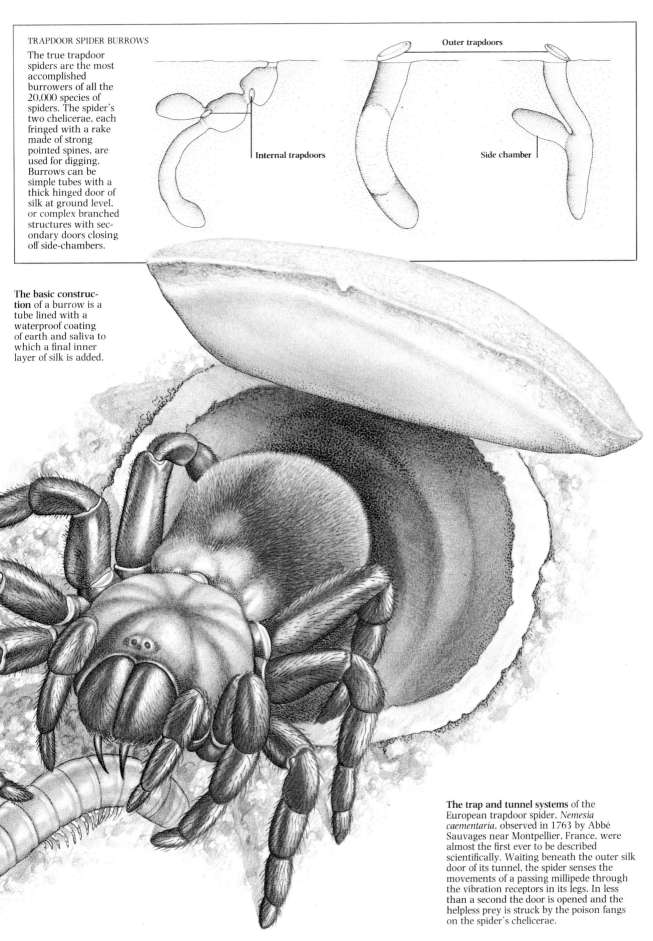

TRAPDOOR SPIDER BURROWS

The true trapdoor spiders are the most accomplished burrowers of all the 20,000 species of spiders. The spider's two chelicerae, each fringed with a rake made of strong pointed spines, are used for digging. Burrows can be simple tubes with a thick hinged door of silk at ground level, or complex branched structures with secondary doors closing off side-chambers.

Outer trapdoors

Internal trapdoors

Side chamber

The basic construction of a burrow is a tube lined with a waterproof coating of earth and saliva to which a final inner layer of silk is added.

The trap and tunnel systems of the European trapdoor spider, *Nemesia caementaria*, observed in 1763 by Abbé Sauvages near Montpellier, France, were almost the first ever to be described scientifically. Waiting beneath the outer silk door of its tunnel, the spider senses the movements of a passing millipede through the vibration receptors in its legs. In less than a second the door is opened and the helpless prey is struck by the poison fangs on the spider's chelicerae.

43

CAPTURE BY TRAPS
The Web Spider

The spider's web is one of the few traps built by a predator. In the spider's abdomen there are glands that contain silk proteins. As these emerge through the fine nozzles of spinnerets at the rear of the body they solidify to form silk fibres that make a functioning web. Some spiders construct a new web, or parts of it, daily; others use the same one for long periods.

Only two of the total 25 families of spiders construct the classic "dart-board design" orb webs normally associated with spiders.

Different webs, however, are produced by spiders in other families. For example, *Atypus* lines a burrow in the soil with a silk tube. Above ground, instead of making a trap door, it continues the silk lining to produce a web in the shape of a closed tube. When insects, worms or wood lice crawl across this tubular part of the web, *Atypus*, lying inside, strikes with its cheliceral fangs through the silk and drags the

prey into the silk tube to consume it.

Linyphiid spiders build a horizontal silk platform web; above it, they construct a three-dimensional superstructure of criss-crossing silk lines attached to nearby vegetation. The spider then hangs from the underside of the platform web. Once any flying or jumping insect hits the superstructure threads, it falls on to the platform and is doomed: the linyphiid strikes and pulls its quarry through to the underside of the web in an instant.

The best-known orb web builders are the argiopid spiders. Typical of the family is the garden spider, *Araneus diadematus*, commonly seen in webs 2 feet (60 centimetres) across and built in low, bushy vegetation. As in all orb webs, strong guying strands of silk are attached to a number of points on surrounding plants to make a circular outer frame. Within this frame, radial "spoke" threads of non-sticky silk form a star-shaped series of supports for a

spiral of adhesive silk that actually traps flying insects.

Once the web is made, the spider sits either at its centre or in a nearby refuge linked to the centre by a communicating strand of silk. The orb web spiders have vibration-sensitive receptor organs in their legs which alert them to the slightest movements of the web. A trapped prey produces a particular pattern of vibrations in the web structure which probably tells the spider about its position and size. If the victim is judged to be catchable, the spider rushes to it, buries its fangs in the prey's body and covers it in silk to prevent its escape. The spider bites the insect again and then carries it off to eat.

The orb web remains the only animal-manufactured trap which everyone must have seen. There are other less-known examples of traps such as the luminous, sticky "fishing lines" made by the cave-dwelling fungus gnat larva of New Zealand for catching prey.

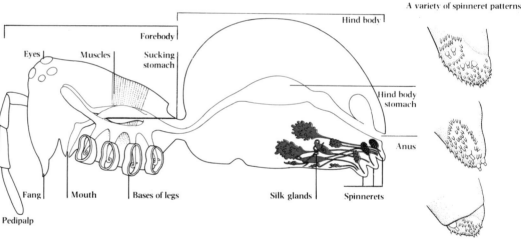

The spider is made up of two parts: a forebody and a hind body. Inside the forebody are the brain and parts of the gut, including a sucking stomach which pumps liquid food into the hind portion of the gut. The hind body contains a digestive gland, reproductive organs and organs connected with silk-making. Liquid silk from the silk glands is forced through spinnerets.

A variety of spinneret patterns

THE CONSTRUCTION OF A WEB

The framework of the web is made with strong non-sticky silk and the catching spiral with sticky (viscid) silk. First, the frame surround and spokes are placed in position with strong guy threads to surrounding vegetation. A few spirals at the centre hold the spokes in place while the spider travels outwards spinning a widely-spaced temporary spiral. With everything locked in place, the spider moves inwards spinning the viscid spiral and removing the temporary spiral.

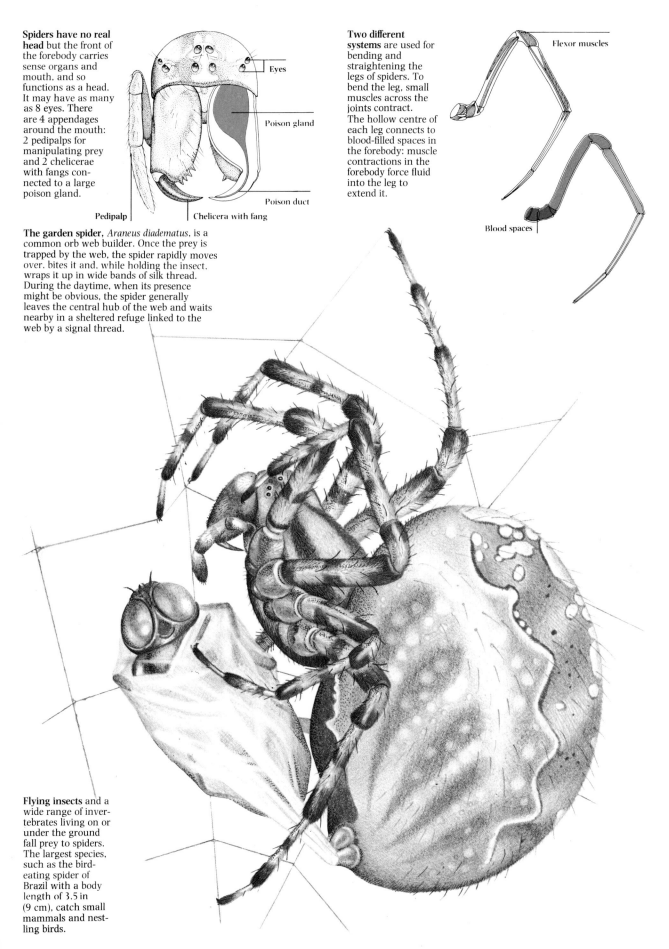

Spiders have no real head but the front of the forebody carries sense organs and mouth, and so functions as a head. It may have as many as 8 eyes. There are 4 appendages around the mouth: 2 pedipalps for manipulating prey and 2 chelicerae with fangs connected to a large poison gland.

Eyes

Poison gland

Poison duct

Pedipalp

Chelicera with fang

Two different systems are used for bending and straightening the legs of spiders. To bend the leg, small muscles across the joints contract. The hollow centre of each leg connects to blood-filled spaces in the forebody: muscle contractions in the forebody force fluid into the leg to extend it.

Flexor muscles

Blood spaces

The garden spider, *Araneus diadematus*, is a common orb web builder. Once the prey is trapped by the web, the spider rapidly moves over, bites it and, while holding the insect, wraps it up in wide bands of silk thread. During the daytime, when its presence might be obvious, the spider generally leaves the central hub of the web and waits nearby in a sheltered refuge linked to the web by a signal thread.

Flying insects and a wide range of invertebrates living on or under the ground fall prey to spiders. The largest species, such as the bird-eating spider of Brazil with a body length of 3.5 in (9 cm), catch small mammals and nestling birds.

The Angler Fish

The shape of the angler fish, unlike that of most fish which are compressed from side to side, has been drastically modified to produce a superbly efficient design for one form of underwater predation. Its highly flattened twenty-four-inch-wide head and thin tail make this unusual fish look somewhat like a tennis racquet.

An extremely wide mouth is the most remarkable feature of the adult angler. Fringed with long, curved, backward-pointing teeth and similar sets of teeth that extend into the roof of the mouth and on to its floor, this cavernous barbed mouth cavity is a death trap for other fish. Its teeth are not used to hack at prey, but simply to make escape impossible. A fish that ventures too close to the jaws is sucked violently in by the powerful water currents created when the angler's mouth opens. The teeth are even hinged so that they rock back to allow large prey to be forced in, but the only way out of the mouth is down the angler's capacious gullet. If the prey tries to back out again, the teeth snap back into their working position making escape impossible.

A poor swimmer, this almost motionless large fish does not rely on the chance swimming patterns of its prey to survive; instead, it attracts fish to the front of its body by a fishing line. On the top of its head is a thin, flexible spine tipped with a double flap that looks like a tiny fish tail. This spine, movable in all directions, is used to jiggle the flap enticingly in front of the jaws. Once smaller fish are tempted to examine it, the line tip slowly moves until both it and the prey are directly in front of the jaws—and then the trap is sprung.

This technique of predation requires little energy, and so the angler needs only a meagre supply of water for gill respiration and the opening of the gill chamber itself is unusually small for a fish as large as the angler.

Remaining hidden is obviously a crucial part of the angler's hunting method, and, as a result, it is adept at camouflage. A swimming angler drops to the sea bottom and disturbs the sediment under it using pelvic fins located under the head and shaped like small hands. Its body then settles halfway into the substratum. The demarcation between bottom and fish is effectively blurred by a fringe of skin flaps around its mouth and sides, and by stumpy pectoral fins, usually pressed down on to the sea bottom, which eliminate shadows that would normally occur at the point where the head region joins the tail. In addition, the angler's muddy-coloured body is covered with warty bumps and filamentous appendages, making it resemble a seaweed-covered portion of the seabed and thus hard for a potential prey animal to spot until it is too late.

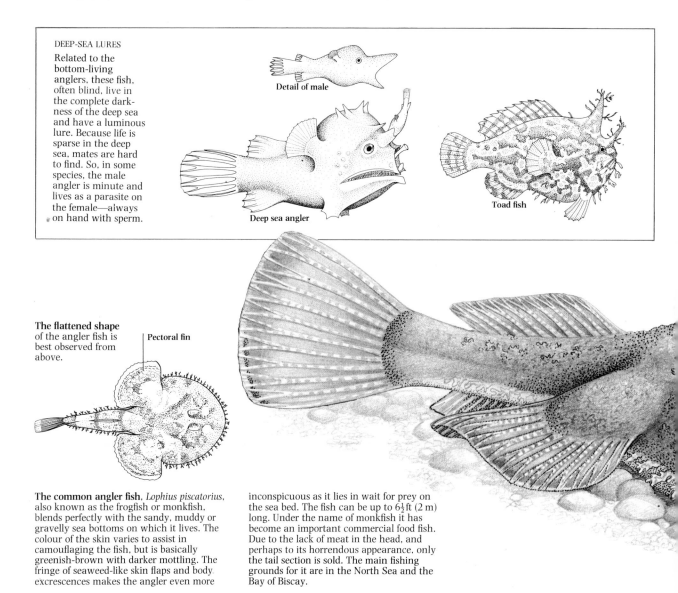

DEEP-SEA LURES
Related to the bottom-living anglers, these fish, often blind, live in the complete darkness of the deep sea and have a luminous lure. Because life is sparse in the deep sea, mates are hard to find. So, in some species, the male angler is minute and lives as a parasite on the female—always on hand with sperm.

Detail of male

Deep sea angler

Toad fish

The flattened shape of the angler fish is best observed from above.

Pectoral fin

The common angler fish, *Lophius piscatorius*, also known as the frogfish or monkfish, blends perfectly with the sandy, muddy or gravelly sea bottoms on which it lives. The colour of the skin varies to assist in camouflaging the fish, but is basically greenish-brown with darker mottling. The fringe of seaweed-like skin flaps and body excrescences makes the angler even more inconspicuous as it lies in wait for prey on the sea bed. The fish can be up to 6½ ft (2 m) long. Under the name of monkfish it has become an important commercial food fish. Due to the lack of meat in the head, and perhaps to its horrendous appearance, only the tail section is sold. The main fishing grounds for it are in the North Sea and the Bay of Biscay.

The teeth on the jaws have a hinge mechanism—each is joined to the jaw at the inner edge by an elastic ligament.

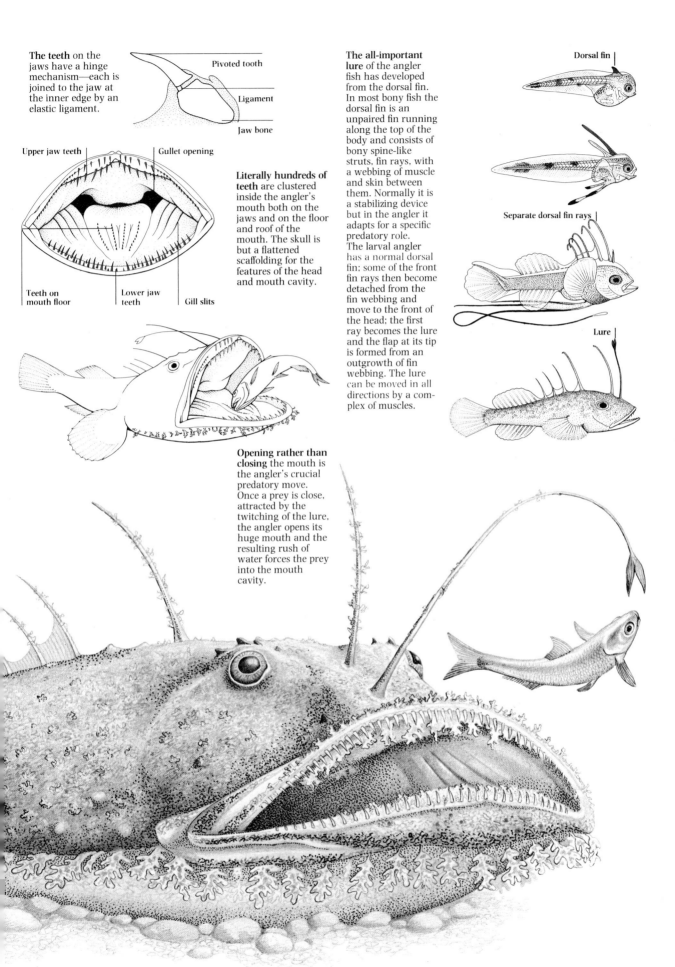

Pivoted tooth

Ligament

Jaw bone

Upper jaw teeth

Gullet opening

Teeth on mouth floor

Lower jaw teeth

Gill slits

Literally hundreds of teeth are clustered inside the angler's mouth both on the jaws and on the floor and roof of the mouth. The skull is but a flattened scaffolding for the features of the head and mouth cavity.

Opening rather than closing the mouth is the angler's crucial predatory move. Once a prey is close, attracted by the twitching of the lure, the angler opens its huge mouth and the resulting rush of water forces the prey into the mouth cavity.

The all-important lure of the angler fish has developed from the dorsal fin. In most bony fish the dorsal fin is an unpaired fin running along the top of the body and consists of bony spine-like struts, fin rays, with a webbing of muscle and skin between them. Normally it is a stabilizing device but in the angler it adapts for a specific predatory role. The larval angler has a normal dorsal fin; some of the front fin rays then become detached from the fin webbing and move to the front of the head; the first ray becomes the lure and the flap at its tip is formed from an outgrowth of fin webbing. The lure can be moved in all directions by a complex of muscles.

Dorsal fin

Separate dorsal fin rays

Lure

47

CAPTURE BY GROUPS: POWER IN NUMBERS

In the age-old battle between predators and their prey some predators have found hunting in groups a successful way to overcome prey defences.

Group hunting has several obvious advantages. First, hunting expertise is pooled—an individual member's strength, such as particularly acute eyesight or hearing, can be used to benefit the whole group. Second, communal "fire power" is far more effective than that of any individual; it allows carnivorous hunting packs to attack much larger prey. Third, co-operative hunting enables the group to use completely new techniques of capturing prey. For example, one lioness cannot be simultaneously a "beater" and an ambusher, but if six females in a pride work together, division of labour becomes possible. One wolf cannot encircle a moose but a pack can.

On the other hand, group hunting does pose some serious problems. The two most important are of a genetic and social nature.

Cooperation in a hunting group implies altruism, which means that for the group to operate as an efficient unit there must be approximately equal sharing of hunt proceeds. The hunting dog that arrives last at the kill because it headed off a zebra must still get a reasonable share of the meat. In a well-regulated hunting dog pack this is precisely what happens. For an individual to give up food that it could theoretically eat itself, however, seems to run counter to the belief that every animal is supposed to fight to the utmost for its own maximal reproductive success. But in a group such as a hunting dog pack, group selection makes altruism a tenable evolutionary strategy. By helping close relatives, an individual hunter is helping other animals with similar genetic makeups. Consequently, most close-knit, highly organized hunting groups that exhibit altruism are extended family groups of one type or another.

Cooperative group hunting suggests an overall complex social life to attain the effectiveness needed in this type of hunting. Usually such societies can operate only if there is subtle and continuous communication between the members of the pack or group.

Both these genetic and social considerations enable scientists to predict the types of animals that are likely to practise group hunting. Social insects and mammals which live in complex societies made up of closely related individuals provide the best examples.

Social insects such as wasps, bees and ants exhibit some of the most extreme characteristics of communal hunters. For them, group sharing of food is complete—each individual will regurgitate food on demand for another member. Among some colonial invertebrates, such as the Portuguese man-of-war, the division of labour is highly specialized. This is also true of ants—one species can have many different morphological types of worker which perform separate hunting jobs.

An unusual type of sex discrimination exists (unfertilized eggs develop into males, fertilized ones into females) among wasps, bees and ants. As a result, thousands of sterile female workers in a single wasp or ant nest have a similar genetic composition; this bodes well for group selection.

Soldier Ants

All ants are highly social insects; they form close-knit colonies, each containing thousands of individuals which are the offspring of one queen ant. A colony is centred on an unusual type of sexual organization.

Males develop from unfertilized eggs; once they have mated with one or more queens, they soon die. Females result from fertilized eggs—the majority become sterile workers, but a small number, when fed on special nutrients, develop into egg-laying queens.

In many ant species, the pattern of feeding of the worker grubs induces them to mature into a wide range of types. In some cases, the workers look alike but vary in size; in other cases, their body forms differ greatly.

Most workers that hunt animal prey have a powerful sting at the end of the abdomen. After overcoming their victim, they try to bring it back intact to the nest, where they then divide it among the members. Ants will regurgitate food for one another on demand. The type and size of prey varies, although sometimes an ant will focus on one particular kind. The North American *Leptogerys*, for example, catches mainly wood lice.

Six *Pheidole instabilis* soldiers, below, are eating a captured caterpillar. In this genus, there is a distinct range of worker sizes. The soldiers are distinguished by large heads and huge jaws, which they use for attacking and immobilizing prey.

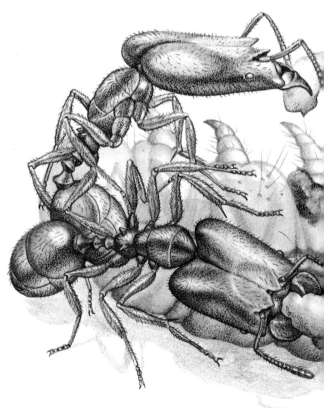

An ant's body is divided into 3 sections: head, thorax and abdomen. The thorax carries 3 pairs of legs and the abdomen the largest portions of the gut. At the rear of the abdomen are the poison glands.

Dufour's gland and a potent retractable sting which together make up a complex poison-producing apparatus.

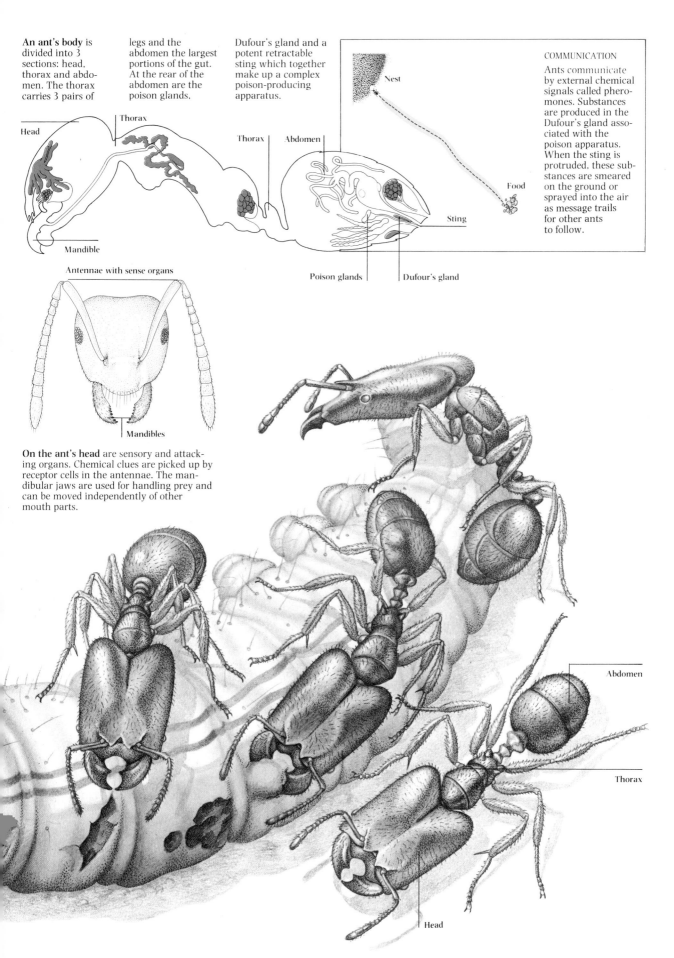

Head

Thorax

Mandible

Thorax | Abdomen

Poison glands | Dufour's gland

Sting

COMMUNICATION

Ants communicate by external chemical signals called pheromones. Substances are produced in the Dufour's gland associated with the poison apparatus. When the sting is protruded, these substances are smeared on the ground or sprayed into the air as message trails for other ants to follow.

Nest

Food

Antennae with sense organs

Mandibles

On the ant's head are sensory and attacking organs. Chemical clues are picked up by receptor cells in the antennae. The mandibular jaws are used for handling prey and can be moved independently of other mouth parts.

Abdomen

Thorax

Head

Hunting Dogs

The hunting dogs of Africa, also known as Cape hunting dogs, have an appalling reputation as communal killers. To many people, the idea of a group of blood-stained dogs disembowelling a breathing zebra is far more horrific than the kill of a lion. In fact, this shared killing results from a high degree of social cooperation within the hunting dog pack.

Found throughout most of Africa south of the Sahara, the long-legged hunting dog is readily identified by its blotchy black, yellow and white coat, large ears and the long white tail tuft which acts as a signal enabling dogs to keep in contact during a long chase through dense cover.

Packs range from twelve to about twenty individuals; each has separate pecking orders for its male and female members. A dominant dog, usually a male, initiates hunting excursions and leads chases. Packs are basically nomadic, preying on gazelles, wildebeest and zebra. It is only when there are

young pups that the group stays in one area with a definite den base for more than a few days.

Hunting dogs spend most of the hot daylight hours in the shade or in dens, sleeping or grooming. At dawn or dusk, one of the dominant dogs becomes active and greets another pack member. Then mass greetings serve as a prelude to a hunting trip by the whole pack, with the exception of young pups and their mothers.

The pack moves off, often in single file, with an experienced dog as leader. At the beginning of the hunt, the dogs trot steadily at about 7 miles per hour, but once prey is sighted, the pack starts running, usually at 30 miles per hour.

Sometimes, a particular animal has been marked out for death at the beginning of the chase. Then the dogs concentrate unerringly on that individual. On other occasions, the pack may pursue several prey together before converging on one animal, or it may take several short rushes at a herd so

that the lame, diseased or slow individuals can be identified.

The chase after a zebra or wildebeest is generally a direct, straight-line test of stamina. Gazelles, however, frequently change direction and thus give straggling pack members a chance to head off the prey. Once the victim is caught, the pack attacks together and eats its way into the prey's soft underparts while it is still alive. The animal may take two minutes to die.

Little aggression is shown between dogs at the kill, as the spoils are divided among all pack members. Late-comers can successfully beg for food from earlier arrivals. Meat is regurgitated for them, as well as for nursing bitches and weaned young dogs.

The cooperation and sharing in hunting activity among pack members arises from their close social identity. This is best shown in the treatment of pups, which are regarded virtually as the property of the pack rather than of the mother dog.

The skull of the hunting dog has special features which relate to its predatory life. Sharp-edged incisor teeth and canine fangs provide the main attacking and biting power. Behind these are a series of molars and premolars used for cutting and chewing hunks of flesh. The other four big teeth, carnassials, act as gristle and bone shears, used when a dog gnaws a bone at the corner of its mouth. The muscles which control both biting and gnawing are the temporalis and masseter muscles. In the hunting dog these provide 90 per cent of its jaw-closing power, so vital in a predatory attack. In a herbivorous prey animal such as a zebra, this figure drops to 60 per cent.

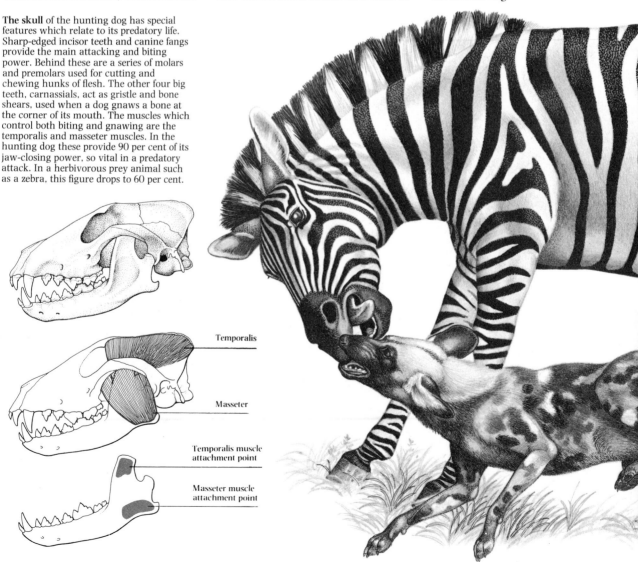

Temporalis

Masseter

Temporalis muscle attachment point

Masseter muscle attachment point

COUNTDOWN TO A KILL

A pack of hunting dogs sights a herd of prey. With heads held low and with a slightly crouching stance, they start to move slowly towards the herd. Once within about 50 yards (45 metres) of their prey the dogs begin to run at top speed, 30–35 mph. Dogs may begin by pursuing different prey but eventually converge on one animal for the kill.

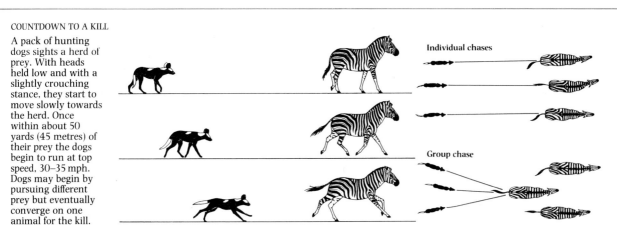

Individual chases

Group chase

At the end of a chase one of the leading dogs in the pack gets a first unyielding hold on the zebra prey. This dog has used the soft sensitive skin of the zebra's upper lip as a vulnerable place. Encumbered and slowed down, the zebra is now an easier target for the rest of the dogs.

Thomson's gazelle | Grant's gazelle | Zebra | Wildebeest

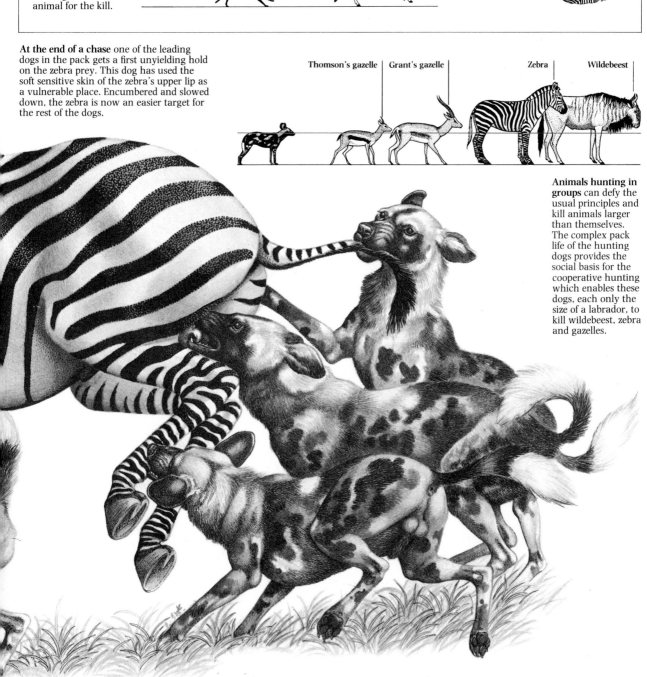

Animals hunting in groups can defy the usual principles and kill animals larger than themselves. The complex pack life of the hunting dogs provides the social basis for the cooperative hunting which enables these dogs, each only the size of a labrador, to kill wildebeest, zebra and gazelles.

Piranha Fish

For centuries travellers to South America have brought back gruesome stories about fish with teeth that can strip the flesh from any animal foolish enough to try to cross the rivers that are their home. Given the name piranha, which means tooth fish, by South American Indians, these much-feared carnivorous predators are restricted to tropical Central and South America.

Piranhas are characins, a group of fish whose primitive forms feed on insects and water plants. Several species, however, live on fruit that drops from trees overhanging rivers. It is from these characins that the flesh-eating piranhas are thought to have evolved.

Although not particularly large, piranhas usually swim in shoals, which can be hundreds or even thousands strong. In such numbers they present a formidable hunting group and it is the presence of the fish in such large quantities that enables them to destroy much larger creatures.

In the four dangerous species of South American piranhas, the upper and lower jaws are both equipped with a single row of triangular, extremely sharp-edged teeth. The set in one jaw fits exactly into the spaces between the set in the other jaw. The mouth looks and behaves like a bear trap made of triangular razor blades. When a prey is attacked, powerful jaw muscles close the piranha's mouth so that the teeth can quickly and efficiently chop bite-sized portions of flesh from its carcass. There is a well-authenticated report of a shoal of piranha reducing a capybara weighing 100 pounds (45 kilogrammes) to a skeleton in less than a minute.

These deadly species of piranha (genus *Serrasalmus*) are distinguished by a flattened muscular body with a blunt front end that carries a forward-thrusting lower jaw. A large bony crest on the top of the skull gives a tough edge to the body. On the underside, there is a keel-like strengthening that bears a row of sharp, backward-pointing spines.

Serrasalmus nattereri, the red piranha, is the most widespread, found throughout the Orinoco basin in Venezuela, the rivers of Guiana, and the river systems of the Amazon, Parana and Paraguay. *Serrasalmus piraya*, confined to the river Sao Francisco in eastern Brazil and one of the most dangerous forms, is the largest of the group, sometimes reaching 2 feet (60 centimetres) in length.

The opinions of experts differ as to the danger posed to man by these notorious fish. Although there are rivers where piranhas are common and people swim with apparent impunity, there are also verified stories of their causing human fatalities. So certain piranhas must be taken seriously as man eaters.

The slender, **muscular tail section** of the body and the stiff, blade-shaped lobes of the tail fin help drive the piranha rapidly through the water towards its prey.

Tail fin | Adipose fin
Dorsal fin

Anal fin

The unpaired fins are concentrated at the rear of the body, giving a large posterior fin area for rapid swimming.

Primitive plant-eating piranhas have intricate teeth. The larger teeth of the upper jaw close outside the lower teeth and alternate with them. The carnivorous piranhas now have fewer teeth, triangular in shape and razor-sharp. The jaws, right, show the full, primitive set of teeth and those retained in the predatory piranha.

Carnivore's teeth | Upper jaw teeth

Lower jaw teeth

Herbivore's teeth

face view | side view

Carnivore's teeth, face view

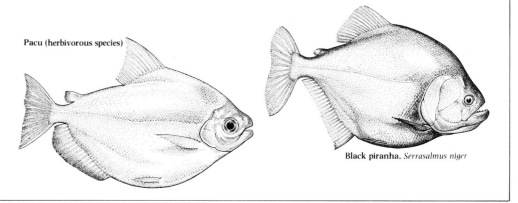

SPECIES OF PIRANHA

The name piranha is applied to about 18 species of characins living in the rivers and lakes of South America. Most are given names indicating whether they are dangerous predators, fish eaters or plant eaters: the blunt-faced forms are known as "piranhas verdadeiras", true or dangerous piranhas.

Pacu (herbivorous species)

Black piranha, *Serrasalmus niger*

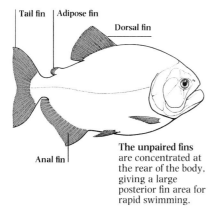

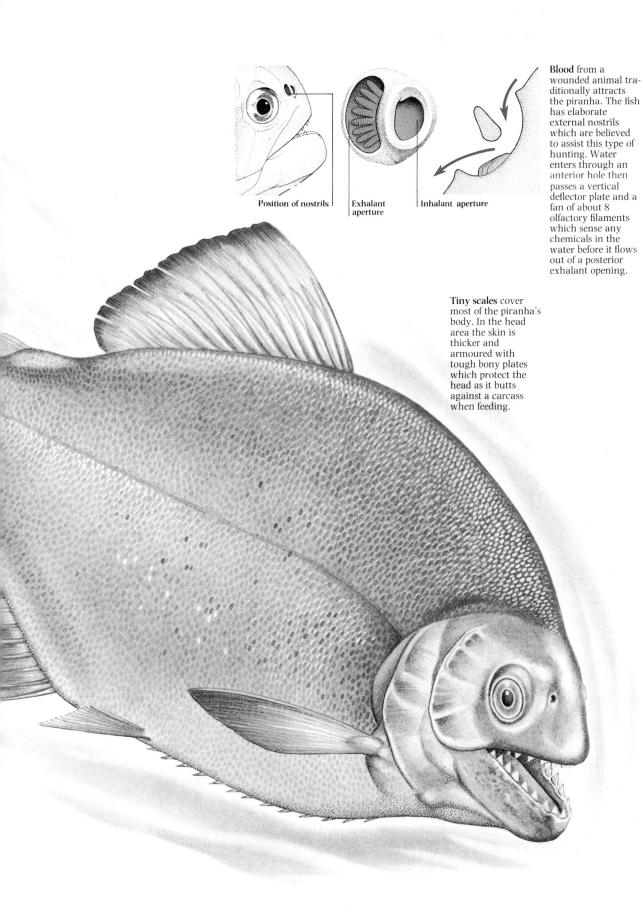

Position of nostrils | Exhalant aperture | Inhalant aperture

Blood from a wounded animal traditionally attracts the piranha. The fish has elaborate external nostrils which are believed to assist this type of hunting. Water enters through an anterior hole then passes a vertical deflector plate and a fan of about 8 olfactory filaments which sense any chemicals in the water before it flows out of a posterior exhalant opening.

Tiny scales cover most of the piranha's body. In the head area the skin is thicker and armoured with tough bony plates which protect the head as it butts against a carcass when feeding.

An attacking red piranha, *Serrasalmus nattereri*, flashes towards its victim. Most piranha stories revolve around this species and there is no doubt that a large group of these fish, each less than 12 in (30 cm) long,
can kill a man when attacking together. The shark-like teeth are extraordinarily hard and sharp: South American tribesmen used to carry piranha jaws with them on hunting trips as an all-purpose tool for
cutting, stripping hide and sharpening weapons. The colours of the red piranha are usually silvery-blue on the sides and brilliant red on the head and belly but they can vary a good deal.

The Brown Pelican

Pelicans are often remembered as the birds whose beaks can hold more than their bellies can. The lower half of their enormous beak can be expanded downwards to form a large distensible pouch for fish capture. No other group of birds uses this technique for catching food.

Pelicans are found only where ecological conditions provide high concentrations of food organisms—such as the Peruvian coast, where the cold Humboldt current from the Antarctic brings nutrient-rich upwelling water to sustain huge populations of zooplankton and the fish that feed on them; the saline lakes of the East African Great Rift Valley, such as Nakuru, which hold vast quantities of algae and plankton sustaining dense fish populations; and the mineral-rich waters of the Danube delta.

Their highly specialized adaptation for food capture means, however, that only a few pelican species have evolved. Authorities disagree as to whether there are six, seven or eight.

Among the largest flying birds, pelicans range from just over 4 feet (1.25 metres) to 6 feet (1.8 metres) in length, depending on the species. All are adapted for an aquatic, fish-eating life. Their most characteristic feature, a huge beak, is long, straight and flattened from above downwards. A medial ridge on the upper bill ends in a down-curving hook that stops fish escaping from the front of the beak. A large gular pouch hangs from the lower bill—empty, it contracts to a ridged fold of skin beneath the beak, but it becomes enormously distended when filled with water and fish.

These gregarious birds breed, fly and feed together in very large groups. The normal feeding strategy of a surface swimming pelican involves underwater sweeps with the open bill, during which the two halves of the lower beak bow outwards so that the catching area is increased. Heavy birds, pelicans rise from the water with difficulty but once up they are strong fliers and often fly in formation. Each bird, except the leader, rides on the upward vortex of disturbed air from the wing tip of the bird in front, thus reducing the amount of energy needed to stay in the air. The leader constantly changes.

The brown pelican, *Pelecanus occidentalis*, with its dark, chocolate-brown feathers, is the smallest species. Well known on the Peruvian coast, brown pelicans fly together in large flocks, identify a region just offshore that abounds in fish, then plunge-dive, catching their prey like gannets do.

The white pelican, *Pelecanus onocrotalus*, of central and southern Asia and the lakes of Africa, has a more refined group hunting method. In shallow water, these pelicans gather together in groups of between six and ten birds in a horseshoe formation. Their beaks plunge forward and downward into the water in unison, to present any fish unfortunate enough to be in the ring with an almost unbroken corral of sweep-net beaks.

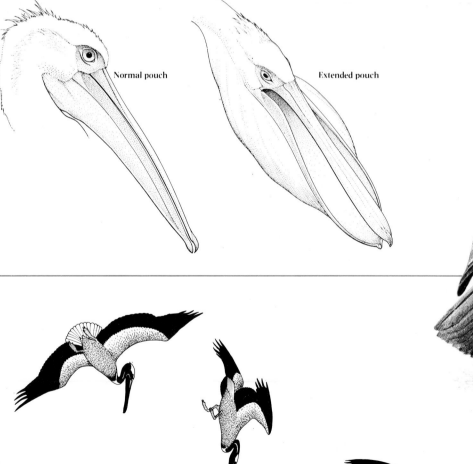

The huge beak with its attached pouch is the pelican's vital food-gathering tool. As the pelican's open beak is pushed under water, the pressure of the water expands the pouch and bows the lower bill into the fishing shape. The pelican brings its head up out of the water with the pouch full; as the bird tilts its head up, the pouch contracts and water escapes from the beak but any fish are retained. The pouch is not a filter; it serves only to take in a large volume of water, and fish, at one time.

Normal pouch

Extended pouch

DIVING

The American brown pelican dives from the air to catch its prey from as high as 50 ft (15.2 m). It is believed to dive with the head held back and the neck curved in an "S" shape as when in flight. This means that some of the initial impact is taken by the front of the body. Air sacs in the skin cushion this impact. The wings are thrown back against the body for a more streamlined outline.

The beak pouch is not a storage space for fish but simply a scoop. The articulations of the neck vertebrae enable the neck to bend forward into an "S" shape in flight or, in the case of the brown pelican, when diving.

Most small fish and swimming invertebrates can be caught with the pelican's pouch. The brown pelican of South America feeds on the dense offshore shoals of anchovies.

The group hunting technique of the white pelican enables the birds to herd together large quantities of fish and facilitates the scoop-beak fishing method. Groups of 6 to 10 birds gather in shallow water. They swim in horseshoe formation closing almost to a complete circle to trap the fish. In balletic unison they plunge their beaks forward into the water.

Fish

Pelicans

The pelican's legs are short and powerfully-muscled. The 4 toes are connected by tough webbing which aids swimming and take-off from the water surface.

The brown pelican, *Pelecanus occidentalis*, is the smallest of all pelicans. The body is dark brown except for splashes of yellow on the forehead and base of the neck and a striking white stripe down each side of the neck. It is the only plunge-diving pelican and one of the most important guano producers of the Peruvian coast. Brown pelicans also live as far north as Baja California and have spread through the West Indies to the southern states of the USA.

CAPTURE BY SPEED: VICTORY TO THE SWIFT

It is no accident of nature that the fastest-flying bird, the fastest-running mammal and the fastest fish in the sea are all predators; the spine-tailed swift, the cheetah and the sailfish all rely on speed to catch their prey. In the constant see-saw battle for success between hunters and the hunted, victory is to the speedy.

To maximize their chances of a kill, predators can be stronger or faster than their prey and preferably both. The way speed is used varies with hunting technique. Neither a mongoose nor a snake moves particularly fast across the ground but both have lightning-quick reactions and can make rapid strikes with their heads in killing attempts. As the snake's neck straightens, it can propel its fanged head forward faster than most small mammals can jump out of range. The rapid head lunges of insect-eating birds unerringly peck on target before their prey can react. Other birds, like the gannet and peregrine falcon, plummet on their prey swiftly and unseen from a great height.

Predators which achieve total body speed do so either by sustained fast movement or by a high-acceleration burst just before the prey is captured. The marlin and the hunting dog use the stamina method. In hunting fish and squid, marlins can keep up high swimming speeds for long periods, while hunting dogs choose a victim from a herd of herbivores such as gazelles or zebras then, as a pack, maintain the chase at 30–35 miles per hour, sometimes for as long as an hour.

The high-acceleration specialists are the hunters that spend much of their lives immobile or in slow searching or stalking. Only at the last, perfectly timed instant do they use a supreme burst of speed to overwhelm their prey. Many of the animals which use this hunting technique have bodies modified for the task. Both marine and freshwater fish may have thin, streamlined shapes which cut down drag in the water. High thrust in swimming is supplied by tall, slim tail fins, plus a large fin near the rear of the body. The sea-going sailfish, which can travel short distances at 60–70 miles per hour, has a tail fin shaped like a vast crescent moon.

The cheetah holds pride of place amongst sprinting land predators. Searching for likely prey—usually the swiftest herbivores such as gazelles—at a stealthy walk, it makes a massive 70-mile-an-hour burst at the last moment, and never for more than a few hundred yards. If it does not achieve the kill then, the cheetah will immediately give up the chase, let the prey escape and await another opportunity.

Although not usually thought of as a sprinter, the African lion also depends on a short, rapid final chase for hunting success. Recent studies of the lion and its prey have revealed that the lion's top speed—45 feet (13.6 metres) per second—is less than that of its common prey, the wildebeest, zebra and Thomson's gazelle. The lion's crucial extra factor is its acceleration. By virtue of its massively muscled legs it can reach its top speed a few seconds faster than its prey. For these tiny but vital periods a lion can gain on all three prey species. The distance between lion and prey when the chase begins is thus critical in deciding how it will end. A lion can increase its chances even more if it is already moving when the prey starts to run—a head start for a kill.

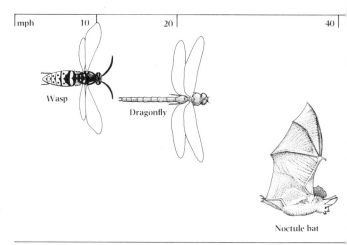

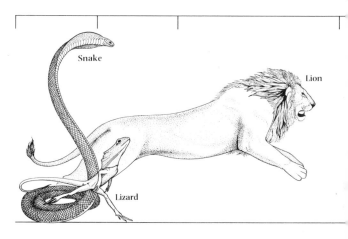

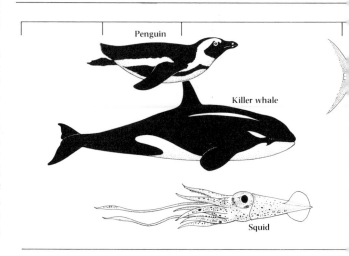

Fast Movers

To move exceptionally fast by water, air or land a predator must have efficient propulsion machinery. For speed in air or water, the predator must overcome the drag its movement causes, so streamlining is crucial. Underwater speed can be achieved in several ways. Large, rear tail vanes are common and fish may also have vertical tail vanes. Dolphins and whales have horizontal fin-like flukes and seals have rear flippers. The undulations of the tail vanes propel the predator forward.

All fast fliers, like the peregrine falcon, use wings which, in birds, are modified forelimbs. Insect wings are extensions of the body wall. On land, streamlining is much less important than stability on the ground and leg action. Equipment for speed over the ground includes long legs and muscles distributed and connected in such a way as to achieve the most rapid leg movements.

60   80   100   110

Spine-tailed swift

Peregrine falcon

Cheetah

Sailfish

Tuna

Air
Many different types of flying predators use speed to overcome their prey. Insects cannot match the speeds of vertebrate aviators but wasps are credited with moving at 10 mph and dragonflies at 20 mph. Birds are the top speed fliers. The peregrine falcon has been timed at over 80 mph in the attack dive and the spine-tailed swift of Asia has achieved 106 mph in level flight.

Land
Land-based predators without legs are at an obvious disadvantage; snakes travel at 10 mph but the head strike speed is greater. The highest recorded speed for reptiles is 18 mph by a lizard, the six-lined race runner. The large carnivores are better equipped for speed. The renowned cheetah is the fastest land animal but the lion can reach a top speed of 30–35 mph in 3–4 seconds.

Sea
Water is the most difficult medium in which to achieve speed because of its high drag and viscosity. Of the marine mammals, seals can move at about 25 mph and the fastest cetaceans, such as the killer whale, at 30 mph. Penguins, with a maximum speed of 25 mph are the fastest birds; sailfish, marlin and tuna, have top speeds of 40 to 60 mph. Squid shoot through water at 20 to 30 mph.

The Cheetah

The most beautiful of the big cats is, by common consent, the cheetah, *Acinonyx jubatus*. Its hunting strategy is based on sheer unbeatable speed. Timed observations over a few hundred yards suggest that an adult cheetah can streak along at 71 miles per hour, thus making it the fastest-running terrestrial animal.

Its lithe, long-legged "greyhound" appearance, with small head and long tail, has been moulded for speed. Most of the body is covered with short yellow-to-tan fur dotted with neat black spots. This feature has given the big cat its common name. Cheetah is derived from the word "chita" in the Hindi language of India and means spotted one.

Although the cheetah's range extends from India westward to Morocco and southward through Africa to South Africa, its distribution is sparse. Only in the plains of East Africa and in Namibia do relatively dense populations occur. The striped-back form from Rhodesia, the king cheetah, is now thought to be a local variety of the once continent-wide species.

Males and females weigh about 130 pounds (58 kilogrammes), making the cheetah the smallest of the African big cats. The length of the head and body is just over 4 feet (1.25 metres) and that of the tail 2 feet (60 centimetres). Muscular and tipped with white fur, the tail is used to maintain balance during the cheetah's high-speed chases.

According to one ecological survey in Tanzania, the cheetah differed from other large predators in its preferred hunting areas—the grasslands. Hunting dogs split their time between grassland and open woodland, as did lion prides. The leopard was rarely seen hunting in the open grassy plains and carried out most of its hunting in the dense woodlands. In areas such as Nairobi National Park in Kenya, however, the cheetah spends some time in all three habitats.

In India, it preys on axis deer and black buck, but in Africa it prefers Grant's and Thomson's gazelles, impala and kongoni. Studies of small hunting groups of cheetahs have shown that they will attack animals weighing between 10 pounds (4.5 kilogrammes) and 600 pounds (272 kilogrammes), with an average weight of 112 pounds (50 kilogrammes).

After a short, rapid chase, a hunting cheetah knocks over its fleeing prey. Once the victim is grounded, the cheetah bites at the throat from the underside and holds on. Examination of prey after kills strongly suggests that death is by suffocation—the bite blocks the windpipe; it does not break the neck or pierce vital blood vessels.

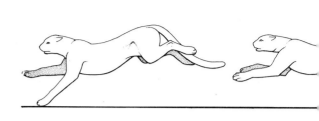

THE RUNNING SEQUENCE

Twice in its sequence of running movements, the cheetah's whole body is off the ground: once with all four legs extended and once with all four bunched under. The distance between two exactly similar points of contact of a foot can be an incredible 23 ft (7 m).

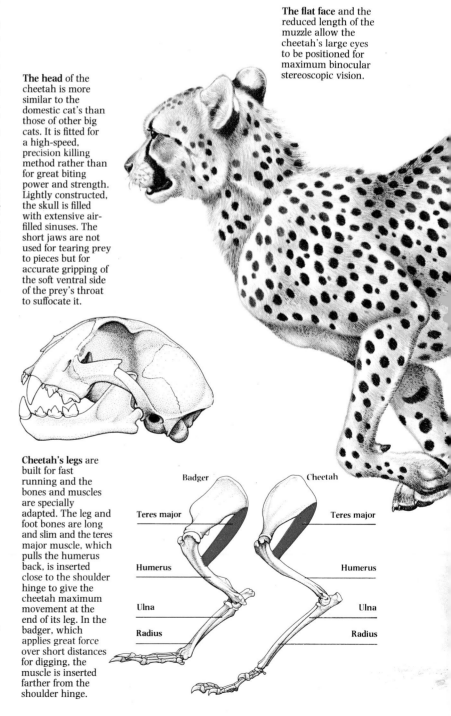

The flat face and the reduced length of the muzzle allow the cheetah's large eyes to be positioned for maximum binocular stereoscopic vision.

The head of the cheetah is more similar to the domestic cat's than those of other big cats. It is fitted for a high-speed, precision killing method rather than for great biting power and strength. Lightly constructed, the skull is filled with extensive air-filled sinuses. The short jaws are not used for tearing prey to pieces but for accurate gripping of the soft ventral side of the prey's throat to suffocate it.

Cheetah's legs are built for fast running and the bones and muscles are specially adapted. The leg and foot bones are long and slim and the teres major muscle, which pulls the humerus back, is inserted close to the shoulder hinge to give the cheetah maximum movement at the end of its leg. In the badger, which applies great force over short distances for digging, the muscle is inserted farther from the shoulder hinge.

Badger

Cheetah

Teres major

Teres major

Humerus

Humerus

Ulna

Ulna

Radius

Radius

**The extreme
flexibility** of the
cheetah's skeletal
framework makes
both its stride length
and its speed
possible. If the back-
bone were rigid, and
the pectoral and
pelvic girdles firmly
attached to the
spine, the cheetah
could not move at 70
mph. In fact, the
backbone curves up
and down alter-
nately during
running and the
limb girdles swivel in
relation to it: at
maximum foreleg
extension the back-
bone is curved
downwards and the
pectoral girdle,
specifically the
shoulder blades,
rocks forwards.

The feet are unusual
in having partially
non-retractile claws,
most probably
because of feeble
development of the
ligaments holding
them back. The
claws, blunt and
short with wear,
help keep a good
grip on the ground
when the cheetah is
running at speed.

The length and size
of the cheetah's tail
are certainly con-
nected with its use
as a balancing
"rudder" to stabilize
high-speed sprints.
The white tip is a
signalling device to
help young cheetahs
keep with their
mother.

The world's fastest land animal, the
cheetah has also had the longest hunting
association with man of any animal with
the exception of the dog—the ancient
Egyptians used cheetahs for coursing game
and in sixteenth-century India, the Emperor
Akbar kept a thousand trained cheetahs for
hunting. In the wild, cheetahs will hunt
alone but also have a significant social life.
Groups of male cheetahs sometimes hunt
together and family groups of mother and
offspring stay together for many months.
Mothers even give their young hunting
lessons: they bring back live prey and
release it for the cubs to chase.

The North Atlantic Gannet

Gannets and boobies are birds of the fertile, shallow waters of the continental shelves where their chief prey—fish and squid—abound.

Man knows the gannets best from their vast colonial nesting areas on desolate coasts, headlands and remote marine islands. The tightly-packed colonies can contain tens of thousands of breeding birds; each nest is only the distance of an outstretched beak from its neighbour.

There are nine known species of these oceanic birds in the family Sulidae. The three temperate-to-cold-water species are called gannets; the six tropical species have acquired the derogatory name of booby, apparently because of the ease with which they let sailors kill them at their nesting sites.

The North Atlantic gannet, *Sula bassana*, is named after its ancient breeding colony on Bass Rock, an islet off the east coast of Scotland. The species is found on both sides of the Atlantic, with a famous colony on the sheer drops of Bonaventure Island in the Gulf of St Lawrence. This graceful sea bird, 3 feet (90 centimetres) long with a wingspan of 6 feet (1.8 metres), has a pearly white plumage, black wing tips and a sulphur-yellow crown. The other two species of gannet are the Cape gannet, *Sula capensis*, which is very similar to the North Atlantic gannet but with a black tail, and the Australian gannet, *Sula serrator*.

Gannets and boobies have specific adaptations for life as marine predators: their totally webbed feet make them efficient swimmers; two salt glands above the eye orbits enable them to feed on salty fish and drink sea water without having to excrete large volumes of urine to remove excess salt. The glands produce a very concentrated salt solution which enters the roof of the mouth via the internal nostril openings and then drips away from the beak.

On their long, thin wings gannets spend most of the year flying over many miles of open sea. Upcurrents are vital for energy-saving, soaring flight and they make use of updraughts, which are caused by winds hitting cliffs or are found rippling in a sequence downwind of an island.

Their long-distance flights are search operations for fish or squid near the surface. Once these are sighted, the gannet then performs a predatory manoeuvre that must be one of the most stirring sights in the animal world. It changes from a straight-winged glider to a swept-wing dive bomber and plunges almost vertically at its prey.

Maximum speed is achieved by the wings angling farther and farther back as the plunge dive progresses. The tapering beak receives the initial impact from the sea, but the front of the skull contains resilient air-filled spaces crossed by strengthening struts of bone. The body skin also has air-filled shock absorbers. The gannet is only under water for a few seconds before it pops buoyantly back to the surface, usually with quarry in its beak.

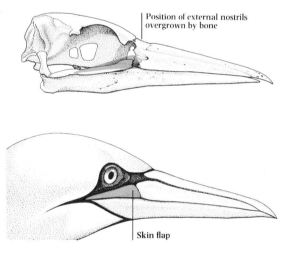

Position of external nostrils overgrown by bone

Skin flap

The dagger-like beak is the gannet's distinctive feature, but its nostrils are the most startling modification. With normal nostrils, water would be forced into the nasal cavities at high pressure when diving, so these are overgrown with a bony flap. The adult breathes via openings in the angle of the beak.

The webbed feet, used for swimming, are also invaluable in the air. Drooped down suddenly, they act as air brakes for slowing down forward flight and for controlling descending glides. When landing, the feet are held out to lessen speed and impact.

THE DIVING GANNET

The gannet catches all of its food in the sea with one of the most spectacular feeding methods of any animal. Fish and squid prey are sighted, perhaps from a height of 100 ft (30 m) as the gannet soars and glides over the ocean. The flying bird then transforms into a diving projectile and hits the water with its wings swept right back and neck and beak held straight out in front.

Air sac of
neck vertebrae

Trachea (wind-pipe)

Lung

Air sacs

Feather

Epidermis

Dermis of skin

Air space

Muscles

The energy require-ments of a flying bird are huge and suggest the need for extra large lungs to fuel this output. In fact the lungs represent only 2 to 3 per cent of a bird's total body volume—5 per cent of man's. Birds satisfy their oxygen requirements in two ways. First, by an efficient system of air movement in the lungs and second, by an expansion of other air spaces, known as air sacs, which are offshoots of the lungs. Thus the total air space volume of a bird is nearly 20 per cent of the body volume.

Air sacs extend into the subcutaneous zones of the gannet's skin creating air cushions in the deeper layers. This feature, present in few other birds, is believed to provide a shock-absorbing layer to counteract the potentially damaging impact when a gannet hits the water after a hundred-foot dive.

Magnificently impressive in flight, the North Atlantic gannet, *Sula bassana*, has a streamlined body and distinctive, pointed, black-tipped wings. In the early 1970s world population of this species was estimated at 350,000 breeding individuals.

The Bottlenose Dolphin

Millions of years ago dolphins, the small cousins of the mighty whales, led a terrestrial and carnivorous existence. Now these ultra-specialized warm-blooded mammals lead totally marine lives and even the reproductive cycle has been transferred to the sea.

The bottlenose dolphin, *Tursiops truncatus*, a favourite performer in dolphinaria around the world, is probably the best-studied dolphin species. In the United States, it is known as the common porpoise. Most commonly encountered along the Atlantic Coast from New England to Florida, it is also found around the British Isles, in the North Sea, along the west coast of Africa and in the Mediterranean.

The dolphin's mammalian body has been moulded and changed to produce an efficient swimming machine. Its streamlined shape, which tapers fore and aft, helps it cut down water resistance. Its forelimbs, with no visible fingers or claws, have been modified into triangular wing-like flippers, which are used for rapid acrobatic steering. And its hind limbs have totally disappeared.

The most remarkable external specializations in this underwater mammal are the tail and dorsal fins, which have become completely new muscular organs. The tail fins, or flukes, emerge horizontally and beat up and down to propel the dolphin forward. The dorsal fin looks exactly like that of a shark and has the same function—to stabilize the rapidly moving mammal against side-to-side rolling movements.

On top of the dolphin's head, directly above its smallish eyes, is the blowhole—a single circular opening that represents the aquatic mammal's nose opening. Its position enables the dolphin to breathe in while staying in a horizontal position with its mouth under water and to breathe out at the end of a dive.

All these adaptations for life spent permanently at sea help make the dolphins versatile hunters. Usually travelling in schools, most species appear to consume fish and large crustaceans as well as cephalopods such as cuttlefish and squid. They actively hunt prey, using their eyes near the surface and at close range, but mainly relying on their specialized form of ultrasonic sonar. They emit high-frequency "clicks" of sound that are bounced off any fast-moving prey. The returning echoes tell the dolphins about prey type, distance and speed, and enable amazingly accurate interceptions to be made.

Most fish and squid are caught between jaws packed with sharp peg-shaped teeth. Dolphins can plummet at least 70 feet (21 metres) and stay submerged for up to a quarter of an hour powered by one breath. Their lungs are relatively large compared with those of land-living mammals. In addition, their bronchial tubes have a series of valves to stop air from being squeezed out of the lungs as they are squashed by water pressure during the dive.

The skull is modified in two ways. First, the upper surface is distorted by the formation of a blowhole at the top of the head. Second, the jaws have become elongated and armed with about 80 teeth for grasping slippery marine prey.

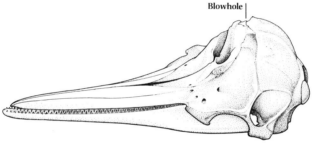
Blowhole

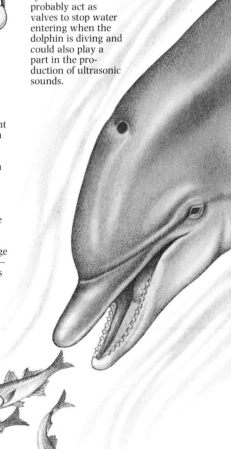
The blowhole on the dolphin's head is a simple circular or crescent-shaped aperture. Inside, a system of sacs probably act as valves to stop water entering when the dolphin is diving and could also play a part in the production of ultrasonic sounds.

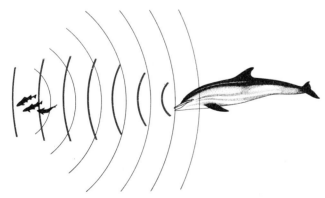

Sonar sounds are beamed out in a narrow cone in front of the dolphin, from a concave bone surface at the front of the skull overlain by a "lens" of fat. Pure tone whistles, thought to be for communication within a school, are one type of sound produced. They are just within the range of human hearing—below 20,000 cycles per second.

Clicks of a much higher frequency are the other type of sonar sound. These are imperceptible to man—up to 269,000 cycles per second—and are used when hunting prey. A dolphin makes about one click a second when gently cruising but up to 500 when chasing prey. The oscilloscope trace, right, records the click sound.

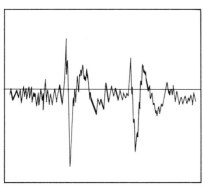

The whale's land-based ancestors, although their exact appearance is not known. probably had skulls similar to that of the modern horse, with nostrils at the front. In the slow evolution of the dolphin, the external nostrils and the bones associated with them have moved back and up the skull to form the blowhole. The premaxilla bone stretches along most of the skull's length and the nasal bone sits on the crown of the skull.

Horse

Nasal bone   Nostrils

Fossil whale

Nasal bone   Nostrils

Dolphin

Nostrils

Nasal bone

The smooth outline of the dolphin's body allows it to achieve impressive hunting speeds: bottlenose dolphins have been timed at 19 mph. The body tapers smoothly at each end, taking it as close as possible to an ideal shape for speed in water. Modifications help streamlining and reduce drag: the head fuses with the trunk; no protruding external ear remains and both penis and nipples are tucked away in body wall slits.

Ideal shape for speed

Fluid filled spaces

The dolphin's skin is also adapted to reduce drag in water. It contains spaces, filled with an oily fluid, that are believed to enable the skin to adjust to and damp out water turbulence.

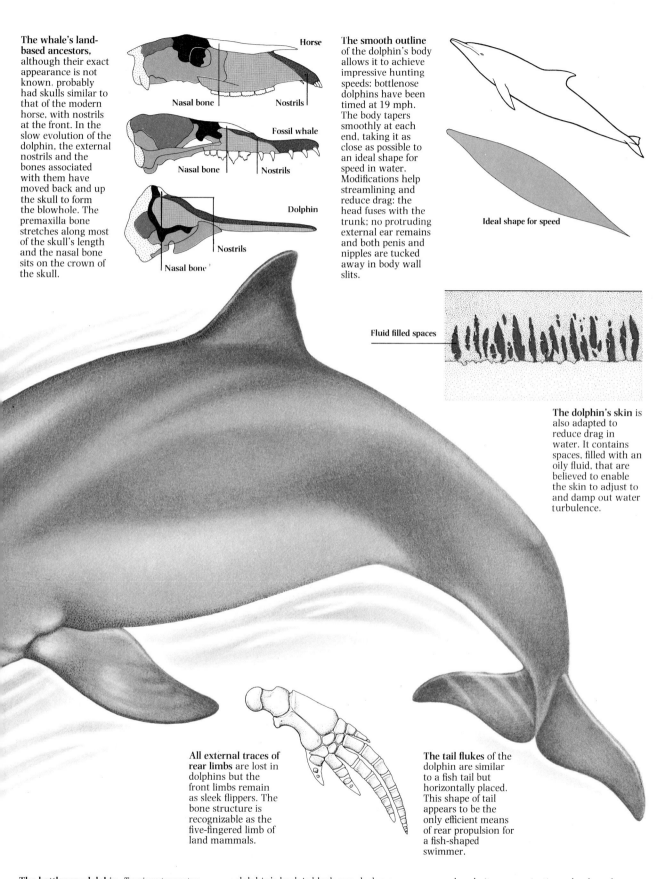

All external traces of rear limbs are lost in dolphins but the front limbs remain as sleek flippers. The bone structure is recognizable as the five-fingered limb of land mammals.

The tail flukes of the dolphin are similar to a fish tail but horizontally placed. This shape of tail appears to be the only efficient means of rear propulsion for a fish-shaped swimmer.

The bottlenose dolphin, *Tursiops truncatus*, eats up to 22 lb (10 kg) of fish a day which it hunts under-water using its sophisticated sonar system. Both male and female dolphins can be up to 12 ft (3.5 m) long and weigh over 400 lb (181 kg). The skin on the dolphin's back is black or a dark grey-brown and the belly is white: this gradation of colour makes the dolphin a less solid-looking object when lit from above and thus more difficult for prey to see. Bottlenose dolphins are social animals; they swim in close-knit communicating schools and show definite evidence of cooperative behaviour—a sick or injured dolphin, for example, will be helped to the surface of the water where it can breathe, by other dolphins.

The Peregrine Falcon

Among the most skilful and acrobatic of flying birds, falcons are unsurpassed as aerial killers, catching birds, mammals and insects on the wing. They are stocky and powerful with tapering, pointed wings and a slim, short tail that make them extremely swift in flight. Most species, however, have retained the ability to kill on the ground when necessary.

The falcon family can be split into three main groups: the forest falcons, which live as predators in the dense forests of Central and South America; the caracaras, which are long-legged, insectivorous or omnivorous birds of the New World with a distinct liking for carrion; and the true falcons, which are found on all continents and are the most adaptable and successful of this family of predatory birds.

The largest of all falcons, the gyrfalcon (*Falco rusticolus*) is almost 2 feet (60 centimetres) long. A magnificent species, with either white or grey feathers and dark spots, this true falcon inhabits Arctic America and similar cold regions of Scandinavia and northern Russia, preying on waders, gulls, rock doves, rabbits and water voles.

The supreme choice of falconers, however, is another true falcon—the peregrine, *Falco peregrinus*. The adult peregrine is 15–19 inches (40–50 centimetres) long; the upper part of the body

is a slaty blue-grey and the underside is lighter with a dark bar pattern. Widespread throughout the world, it is now split into 17 recognizable local subspecies. It is adaptable in terms of habitat, and actually breeds in towns, apparently attracted by the large numbers of domestic pigeons to be found there. In Britain, though, it is a bird of wild and desolate countryside. Its speed and manoeuvrability during a hunting flight are probably unequalled by any other bird.

In the sky the hunting peregrine presents a crossbow-like silhouette. Its characteristic hunting method is to circle high over its victim, often a rock dove, before making a near-vertical dive, or "stoop", striking the prey from above with outstretched talons. It then follows its plummeting victim to the ground, or forces it down if it is not completely disabled. The peregrine plucks at the carcass before finally shredding and eating the flesh with its hooked beak. If the direct, 80-mile-per-hour stoop fails, the peregrine will chase a bird in the air with dramatic acrobatic changes of direction. Peregrines can reach 60 miles per hour in level flight.

This superb falcon kills almost any flying prey smaller than itself. More than a hundred different bird species have been recorded as prey items as well as certain mammals, frogs and insects.

Red-thighed falconet

Common caracara

Collared forest falcon

Hobby

The basic falcon design is adapted in different species for efficient predation. Tiny, forest-dwelling falconets of the tropics are only 6 in (15 cm) long and feed on insects. The long-legged caracaras of South and Central America are scavengers, while forest falcons, in South America, hunt ground-living prey. Hobbies are long-winged, fast-flying falcons and catch their food in the air.

The kestrel, also known as the sparrowhawk, hunts for prey while hovering on gently fanning wings, its head motionless, over one spot of ground.

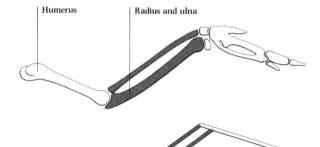

Humerus	Radius and ulna

A double bone, the radius and ulna, connects the first wing bone, the humerus, to the bones carrying the wing-tip feathers. When the humerus folds towards the radius and ulna the wing tip angles back, facilitating the dive.

The dive of an attacking peregrine falcon has been reliably timed at 82 mph. As the dive speed builds up, the falcon becomes an increasingly sleek projectile, reducing the drag caused by wings and tail: the wings become progressively swept back with a tapered, pointed outline and the tail assumes a narrower shape.

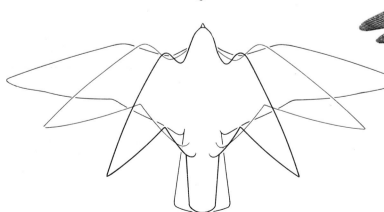

Like a closed fan, the peregrine's tail feathers are folded together in a dive to give a slim, streamlined outline. This shape does not, however, allow much control surface, so the tail opens out to a narrow wedge shape when changing direction or in slower flight.

External nostril

Hooked upper bill

Lower bill

The hooked beak of the peregrine falcon has the overhanging pointed tip to the upper bill so characteristic of carnivorous predatory birds. Prey is usually killed by the strike of the talons rather than the beak but when preparing the carcass for feeding, the falcon employs its bill to pluck out the prey's feathers and rip flesh from its body.

A peregrine falcon ends an attack dive with talons outstretched to strike its prey, a white-throated sparrow. This falcon actually does "peregrinate" over long distances out of the breeding season: two of the North American race have even turned up as far afield as Britain. Sadly, this splendid bird has suffered a disastrous decline in numbers over recent years. This is thought to be due to its ingestion of toxic insecticides, which are concentrated as they pass through the falcon's food chain—insects, insect-eating birds, peregrines.

CAPTURE BY SPEED
The Mongoose

Agility, speed and versatility in the hunt are the talents of the mongoose. This slender, long-bodied carnivore shares, with the civets, genets, linsangs and palm civets, a place in the Viverridae, a diverse mammal family exclusive to the Old World. Although they possess features common to all carnivores, the mongoose and its relatives are opportunist rather than adapted to one particular life-style. It is believed that they—or animals very like them—were the jumping-off point in the evolution of the specialist carnivores such as the cats and hyenas.

Mongoose species can be as big as a domestic cat or as small as a weasel. As predators, they have a deserved reputation as efficient snake killers, but the various species will take any small animals, including beetles, mammals and birds. Eggs are favoured food.

In some parts of the world mongooses have been partially domesticated to keep houses clear of rats, mice and snakes. Man has taken the small Indian mongoose, *Herpestes auropunctatus*, from its natural home in the Indian subcontinent to act as a rodent controller in the West Indies and Hawaiian islands. Intensive investigations of these compulsory immigrants have shown them to be versatile predators. As well as exterminating rats and mice, they hunt spiders, grasshoppers, scorpions, centipedes, snakes, frogs and toads. As far as their household job of rodent killing is concerned, however, these mongooses have mixed success. Being daytime hunters they are adept at preying on brown rats, which feed by day and live in burrows, but have little effect on black rats, which feed by night and nest in trees and among roots.

On the Hawaiian islands, and in other places with large areas of seashore, the Indian mongoose has changed its habits to become a crab eater. It forages along the shore line and even wades in the water to overturn stones in search of its quarry. In Africa's swamplands the marsh mongoose. *Atilax paludinosus*, another specialist, has similar habits, taking frogs, crabs, fish and water snails from its marsh habitat.

Other variations on the mongoose pattern of predation include the banded mongoose, *Mungos mungo*, which eats a mixed diet but will—as it does in Uganda—feed largely on insects, and the white-tailed mongoose, *Ichneumia albicauda*, which also concentrates on insects. The white-tailed mongoose is a nocturnal hunter and often frequents villages and the outskirts of towns to take advantage of the moths and other night-flying insects attracted to the lights of human habitations.

FIGHT TO THE DEATH
The agile, fast-moving mongoose can overcome snakes much larger than itself. Typically, the mongoose darts inside the snake's reach, bites and holds the head and jaw. Once it has this hold, the mongoose may be knocked off its feet and enveloped by the snake's coils but can eventually make a killing bite to the tiring reptile's head.

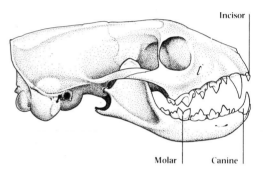

The elongated skull of the mongoose shows the tooth pattern so crucial in predation. Rows of incisor teeth form a cutting edge at the front of both jaws and pointed, protruding canines can clamp on to a snake's head. Strongly developed molars have pointed cusps for crunching insects.

Incisor

Molar   Canine

The banded mongoose is almost entirely insectivorous and often preys on dung beetles and flies living in elephant droppings.

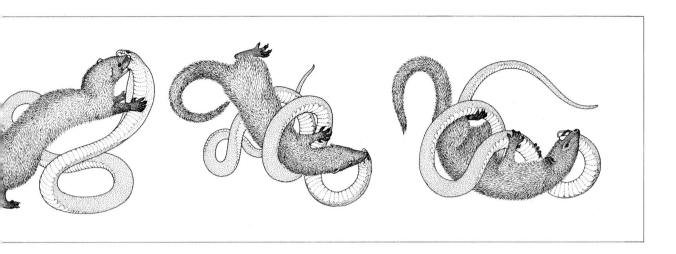

Each paw of the mongoose has 5 digits. The manipulative front paws have strong claws, the toes are separate and the palms generally free of hair.

Strike speed, poison fangs and attention-distracting devices are the snake's weapons in battle. The cobra can expand the area beneath the head to make itself look larger and more threatening and perhaps divert the predator's bite.

In a battle with a mongoose, the cobra nearly always loses—for several reasons. First, the cobra has a standard attack method. It rears up to strike, which allows the mongoose to jump inside its attack radius. Second, the cobra's fangs are fixed and cannot lead the strike as the jaws do not open sufficiently wide; so, while the mongoose may be hit by the snake's head, it has time to avoid the fangs. Third, the fur of the mongoose stands erect while it is fighting, making the animal seem larger and causing most of the snake's strikes to fall short, hitting the fur rather than the skin.

SPECIALIST HUNTERS: EXCLUSIVE TASTES

In the world of predatory animals, becoming an ultra-specialist is apparently an exceedingly dangerous strategy, for such creatures are rare in the normal pattern of life. Far more common are generalist predators that can adapt to different diets and life-styles. Predators, it seems, run a great evolutionary risk of extinction if they come to rely exclusively on an unusual hunting technique or a restricted range of prey species.

Along the path of evolution, extreme specialization and vulnerability move hand in hand. If an animal alters, through a series of slow evolutionary changes, to achieve a supremely efficient, intricate mechanism for capturing and devouring prey, it will almost certainly prosper for a while. But, in the long term, such a commitment bears the seeds of danger, for success is only assured while biological conditions remain unaltered. If the climate or the patterns of competition between other animals change so that the hunting technique becomes untenable or the preferred prey species rare, then suddenly the specialist is left high and dry. And an extreme specialist is unlikely to be able to change quickly enough to adapt to the new circumstances.

Adaptability is, throughout the story of evolution, the key to survival and extreme specializations mean compromising adaptability. If the ability to adapt becomes too diminished, only clearly defined biological circumstances can ensure a highly specialized predator a long evolutionary history. Given this gloomy prognosis, how is it that extreme specialists do exist? How can predators with bizarre equipment like the duckbill of a platypus or the tusks of a walrus manage to survive, and how can the aardvark exist on one prey species?

In the survival of ultraspecialists a predictable food supply is an important factor. If, over increasingly long periods, a particular prey species is plentiful and readily available, it becomes more and more likely that this species can be relied upon as a source of food. In Africa, for example, termites are ubiquitous and extremely numerous insects, and have been so for millions of years. The termites form such a stable food supply that an animal like the aardvark can afford to live a life almost exclusively devoted to capturing and eating them. But if some ecological catastrophe—perhaps caused by man—eliminated termites from the face of Africa overnight the superspecialist aardvark would, almost inevitably, perish to the point of extinction.

In becoming so specialized the aardvark has made the ultimate commitment, for its whole body structure and its behaviour are adapted to termite hunting. Extreme specialization need not, however, involve such drastic moulding, for predatory behaviour can be refined and complicated without the need for massive structural transformations. The Everglade kite, for example, which concentrates its predatory efforts on one particular snail species, is not remarkable anatomically, nor are the bee-eaters and indicator birds that rely almost entirely on bees for their food. If a cataclysm destroyed the world population of bees or the Everglade kite's snail prey overnight, it is far more likely that these three birds, which are specialized in behaviour, not anatomy, could make some sort of adjustment to new types of prey than an aardvark could if deprived of termites.

The Everglade Kit

A beautiful, long-legged bird of prey, named after the marshlands of the southern United States where it was once common, the Everglade kite, *Rostrhamus sociabilis*, is the ultimate specialist in selecting prey. It exists exclusively on a single type of water snail, *Pomacea*.

Near the marshy pools where these snails abound, an Everglade kite will sit on a roadside fence post or some other vantage point to observe its quarry. Late afternoon and evening are the kite's usual hunting hours, for this is when the snails are most active and move on to low-growing vegetation in the pools.

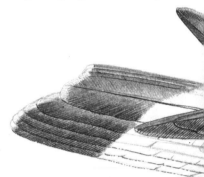

The hunting kite flies slowly over the marshlands searching for prey. When it spots a snail, it hovers briefly, long legs dangling, then drops down to pick up the snail in its talons.

With its long, tapering upper bill, the kite strikes the body of the snail at the instant it begins to emerge from the shell. This puncturing blow appears to damage the nervous coordination and the impaled snail becomes limp and relaxed. A shake of the kite's head finally removes the contents from the shell. The bird then swallows the flesh of the snail allowing the shell to fall to the ground beneath its perch. It never actually pierces the shell.

Flying and hovering over the water until it sees a snail, the kite grasps the prey and takes it back to a favourite perch. There, using dextrous, precise beak movements, the kite extracts the soft, edible contents from the shell and swallows them whole. The shell is discarded. The piles of shells that soon build up beneath each eating post are mute witness to the efficient nature of the bird's hunting technique.

The huge, regularly available water snail populations must be one reason why the Everglade kite can afford to be such a choosy predator. But why this particular prey can sustain an ultra-specialist predator while other, equally common prey species cannot, remains a mystery. Possibly the behavioural trick of removing the contents of the tough snail shell is one that has evolved only rarely, allowing any animal achieving such a faculty to streak ahead of its competitors and afford the luxury of great specialization.

The most severe threat to the survival of the Everglade kite is man, who has not only drained vast areas of marshland, thus robbing the kites of water snails, but also shot many individuals. Large populations do still exist, however, in Central and South America.

Hook-billed kite

Two other American birds are specialists at catching and eating snails. The hook-billed kite, *Chondrohierax uncinatus*, above, uses its bill to winkle out the contents of land-dwelling snails. Another, the limpkin, *Aramus quarauna*, is in direct competition with the Everglade kite as it also feeds on the snail *Pomacea*.

Carrying the snail in a long-toed foot, the kite returns to its feeding post. It perches on one foot while quietly holding the snail in the other. It makes no attempt to remove it by force but waits for the snail to emerge.

The Walrus

The freezing Arctic waters round the north polar ice sheet are the home and hunting ground of the world's one species of walrus—a highly specialized predator related to the seals and sea lions. Its family name, Odobenidae, "those that walk with their teeth", gives the clue to the walrus's offensive weapons, for this giant, sea-going mammal has a pair of huge canine tooth tusks—massively elongated teeth of the upper jaw.

The walrus employs its vast tusks, which may be as long as 3 feet (1 metre) in old bulls, to pull itself on to ice floes, in sexual battles between warring males and, most importantly, as digging tools for exposing prey. Diving in search of their food to over 200 feet (60 metres), walruses seek out clams, other bivalves and sea snails from bottom sand or mud. The tusks plough up the sea-bed sediment, exposing the shelled food which

they recognize in the lightless depths by snout whiskers, the vibrissae. The front surface of the snout is densely arrayed with neat rows of these stiff, quill-like hairs which are extremely sensitive to touch.

Using its non-tusked teeth—four on each side of both upper and lower jaws—the walrus crushes the shells of its prey then swallows the soft animal bodies whole, rejecting the shells. When these crushing teeth first develop in young walruses they are conical, but in older animals they are invariably flattened by many months of grinding down the shells of their prey.

Apart from this usual feeding pattern, walruses will eat much larger prey— seal blubber is sometimes found in their stomachs. There is even a well authenticated account of a walrus eating a narwhal—another tusked mammal.

Walruses live together in family herds

of cows, calves and young bulls consisting of up to 100 individuals. Adult bulls congregate in separate herds, only mixing with females in the mating season from late April to early June. About a week after birth a walrus pup can swim, dive and feed alone but still hitches rides on its mother's back, holding on with its flippers. The walrus can live for more than 30 years if undisturbed by enemies. It was once sought after by commercial hunters for the ivory tusks and blubber oil but is now protected.

The walrus is not usually aggressive to man, but the massive bulk of a male walrus and its formidable tusks make it a frightening adversary when roused. Hunted animals have speared the sides of boats, and polar bears, fearless killers of seals, will baulk at attacking an adult walrus, even on land where the walrus is at its most vulnerable.

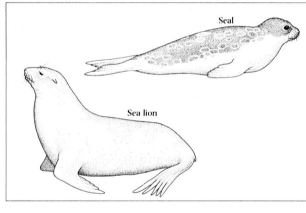

Seal

Sea lion

SEALS AND SEA LIONS
Walruses are related to other marine mammals—the sea lions, or eared seals, and the true earless seals. Walruses are closer to sea lions and, like them, can turn their hind flippers under to move on land. True seals are more committed to marine life and move awkwardly on land.

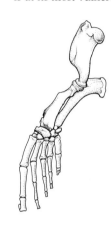

The front flipper of the walrus is a strong oar-like structure developed from the walking leg of its terrestrial ancestors. The flippers are an important means of propulsion in water.

The walrus dives to find its food, sometimes as deep as 300 ft (90 m). It is believed to have similar physiological adaptations to seals, the most important of which is a great reduction in the heart rate while diving under water. Blood flow is reduced and concentrated in essential organs.

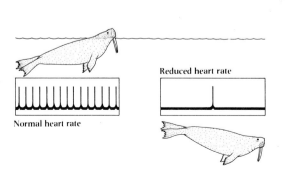

Reduced heart rate

Normal heart rate

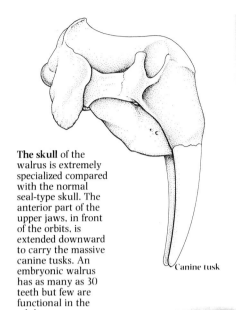

The skull of the walrus is extremely specialized compared with the normal seal-type skull. The anterior part of the upper jaws, in front of the orbits, is extended downward to carry the massive canine tusks. An embryonic walrus has as many as 30 teeth but few are functional in the adult.

Canine tusk

FOOD FINDING

There is little first-hand evidence on the exact method the walrus uses to gather its food but it appears to use the tusks to dig out shellfish from the sea bottom. Clams are the main food but gastropods, annelid worms, crustaceans and starfish are also eaten. Once the food is uncovered, the mobile and pro-trusible lips over-come the problem of getting it into the mouth.

Mollusc prey in bottom sediment

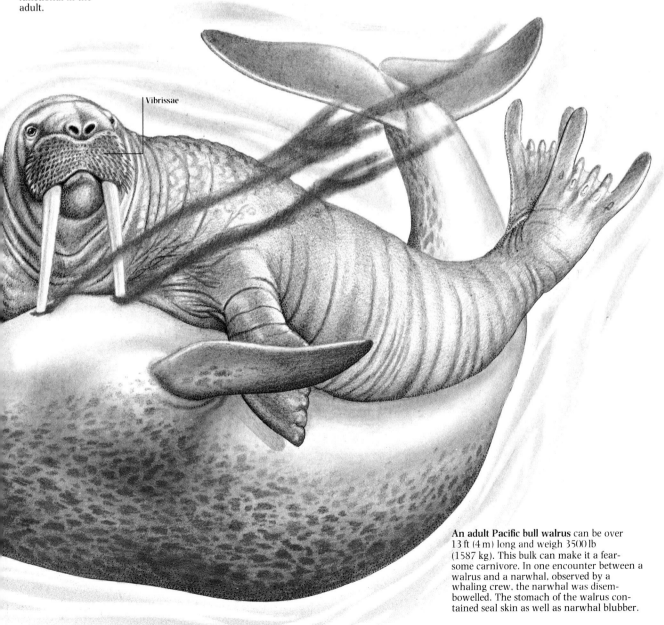

Vibrissae

An adult Pacific bull walrus can be over 13 ft (4 m) long and weigh 3500 lb (1587 kg). This bulk can make it a fear-some carnivore. In one encounter between a walrus and a narwhal, observed by a whaling crew, the narwhal was disem-bowelled. The stomach of the walrus con-tained seal skin as well as narwhal blubber.

The Aardvark

The aardvark is a supreme specialist among predators. It is a singular animal—adapted for digging and built to feed almost exclusively on termites, the minute, nest-building, social, herbivorous insects. Utterly alone on the animal family tree as the sole survivor of a whole order of mammals, the aardvark has a bizarre anatomy consisting largely of extreme specializations which enable it to find and eat enough of these insects to survive.

The old Afrikaans word meaning earth pig is the source of the aardvark's name. Appropriately, a sparse covering of hair gives the aardvark a bald, pig-like appearance and the long, nostrilled snout reinforces this porcine impression. It measures up to 6 feet (1.7 metres) in length overall—the thick, kangaroo-like tail accounts for 2 feet (60 centimetres) of this—and weighs up to 154 pounds (70 kilogrammes).

The powerfully muscled legs that support the aardvark's body are equipped with superbly efficient digging claws—those of the four toes on each foreleg are so massive that they look like hooves. With these claws an aardvark can rip its way into the tough, sun-hardened walls of termite nests, into the ground to expose termite tunnels and into the rotten wood on which termites may feed. The exact digging technique varies, but an aardvark will often squat in front of a conical termite nest on its hind limbs and tail. With the forelimbs thus freed for action it can rapidly open a hole in the side of the nest.

Digging ability is also vital to the aardvark's self-preservation, for it feeds mainly at night, then makes a burrow into which it retreats for protection during daylight hours. Each burrow is 12 to 15 feet (3.6 to 4.5 metres) long with a chamber at the end where the aardvark can turn round. In South Africa its burrowing speed is legendary—by all accounts it can outstrip a team of six men digging with spades. When burrowing fast in defence of its life, the aardvark breaks up the soil with its front claws and pushes this soil back to be kicked aside by the rear feet. An aardvark may, at least temporarily, lie up among its prey by excavating a hole in the centre of a termite nest. Its thick, almost hairless skin, underlaid with very little fat, apparently protects the aardvark from the bites of insects.

Like the anteaters, pangolins and other mammals with a diet similar to its own, the aardvark has a long snout. Its head houses sensory and feeding apparatus highly adapted to a termite-eating existence. The nighttime feeding aardvark relies heavily on its well-developed sense of smell. The bones within the skull on which the tissue (olfactory epithelium) containing odour-receptive cells is spread are more extensive in the aardvark than in any other mammal. The nostrils themselves are protected by bristles.

Of all the aardvark's unusual features, the mouth is the most extraordinary. Its tongue is an adhesive death trap for termites. This highly extensible organ, which can be 18 inches (45 centimetres) long, is covered with sticky saliva, thus ensuring that termites are captured in large numbers at a time. The adult aardvark has no incisors or canines (biting or cutting teeth) and the cheek teeth, which grow unceasingly throughout the animal's life, are unique in construction—they lack roots and enamel and are used to crush prey.

On the route map of evolution the aardvark is at the end of a blind alley, totally committed to living on one of Africa's dominant insect forms. But while termite mounds still tower from Africa's plains it has a secure niche—while man does not interfere.

Unique teeth give the order of which the aardvark is the sole member its name: the Tubulidentata, or tube-toothed ones. Dentine in the cheek teeth is arranged in hexagonal zones around tubular pulp cavities. The teeth have no roots or surface enamel.

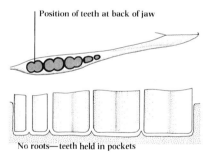

Position of teeth at back of jaw

No roots—teeth held in pockets

Supersensitive ears give the aardvark good warning of any approaching enemy so it can burrow out of sight to escape attack. If cornered, the aardvark will fight with its powerful clawed feet. While digging for termites the ears are folded back to close the ear opening.

The probing snout is protected in two ways as the aardvark throws back soil: the external nostrils can be closed and are defended by a ring of stiff bristles surrounding the muzzle.

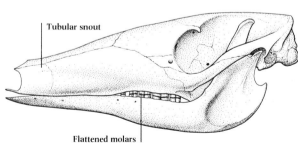

Tubular snout

Flattened molars

The skull is no less unusual than the external appearance of the aardvark's head. Along much of the slim, protruding snout the expanded surfaces of the turbinal bones support a large area of olfactory sense cells: sense of smell is crucial for this nocturnal hunter. The anterior incisor and canine teeth are absent in the adult, allowing the all-important tongue free movement from the small mouth. The unique cheek teeth, specifically adapted for crushing the termite prey, are positioned to the rear of the aardvark's jaws.

Termites are trapped on the aardvark's sticky tongue. The tongue is kept wet by large salivary glands and can extend to 18 in (45 cm).

THE TERMITARIUM
Termite mounds vary in structure but many are as high as 10 ft (3 m). The mound provides some defence for the termites as well as regulating the temperature and humidity inside. The outer wall is almost rock-hard—a sun-baked mixture of soil, plants and termite secretions.

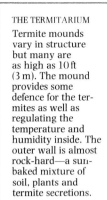

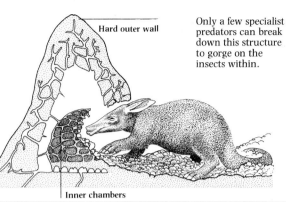

Hard outer wall

Only a few specialist predators can break down this structure to gorge on the insects within.

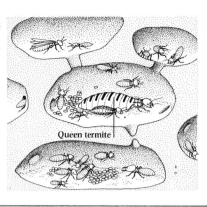

Queen termite

Inner chambers

A comparison of the brain anatomy of a primate, such as a chimpanzee, and the aardvark reveals telling differences. The most important contrasts are the emphasis on regions concerned with the analysis of smell information in the aardvark, principally in the form of a large olfactory lobe, and the relatively small size of the aardvark's cerebral cortex.

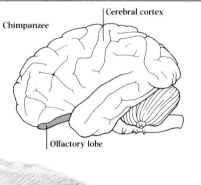

Cerebral cortex

Chimpanzee

Olfactory lobe

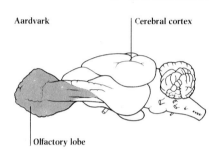

Aardvark

Cerebral cortex

Olfactory lobe

The strong tail of the aardvark serves as a prop when the animal sits on its hind quarters to burrow with its front feet into a termite mound.

The Bee-eater and the Honeyguide

The brilliantly coloured bee-eaters and drab honeyguides, or indicator birds, may look completely different but, as predators, share a dependence on hymenopteran insects—particularly bees—for their food. These groups of birds, both found chiefly in Africa, go about their bee hunting in utterly different yet equally interesting fashions. Bee-eaters are airborne hunters, while honeyguides can exist on a highly specialized diet of bee grubs and beeswax by virtue of mutually beneficial (symbiotic) relationships with other creatures.

The bee-eaters comprise a family of 25 of the world's most gaudy birds—almost every species is splashed with vivid reds, greens, purples and other bright colours. With long, tapering wings and tail feathers they are both buoyant and acrobatic in the air. Nearly always, bees, wasps and a few other insects such as locusts are taken on the wing and gripped near the end of the bee-eater's long, slightly downward-curving beak. The bee-eater carries the prey back to a favoured perching site—a branch, rock, or very often a telephone wire—in its beak. Here the insects are scraped and wiped along the perch (presumably to remove the sting) before the bird devours them.

Most bee-eaters are gregarious, performing exciting mass flights. Most are also colonial nesters. Nests are made at the ends of long, horizontal burrows in river banks or sand cliffs.

Compared with the psychedelic colours of the bee-eaters, the dull browns, greens or greys of the honeyguides make them seem complete nonentities. But their appearance belies a startling pattern of symbiotic relationships. Two honeyguide species, including the aptly named *Indicator indicator*, the greater or black-throated honeyguide, lead both man and the honey badger (ratel) to bees' nests with excited cries. Once man or ratel has demolished the nest—taking the brunt of the bees' stinging attacks in the process—and robbed it of its honey, the honeyguide moves in to devour both bee larvae and the wax of the nest itself. Their thick skin may well protect them from stings. The honeyguides can also catch insect prey in the air on their own account.

The vivid red and green plumage of the carmine bee-eater, *Merops nubicus*, makes it one of the most beautiful of this group of birds. There are large colonies of this species all over tropical Africa. Bee-eaters may be up to 14 in (35.5 cm) long including the characteristic elongated tail feathers.

THE MOBILE PERCH
The carmine bee-eater often makes use of another bird, the kori bustard, as a mobile perch and as a "beater". As the larger bird walks through the grasses of the East African plains, it disturbs insects which fly up and are caught by the bee-eater. It treats the kori's back as a take-off and landing strip as it chases and catches prey.

Honeyguides are related to woodpeckers and barbets. The black-throated species, *Indicator indicator*, is widely distributed throughout tropical and southern Africa in most types of vegetation other than dense forest. As well as feeding on bee larvae, the honeyguide has the unusual ability to digest beeswax. This is made possible by symbiotic bacteria in its gut which helps break down the wax. The guiding habit is thought to have originated with the honey-eating ratel. Man observed this and imitated at some point; now the birds have come to regard man as an equally useful hunting partner.

As young honey-guides would find it hard to cope with a diet of bees and wax, the eggs are laid in other birds' nests. The bird hatches with hooks on its beak, which it uses to destroy the rightful young in the nest and thus get all the food. The hooks drop off when the bird is about 2 weeks old.

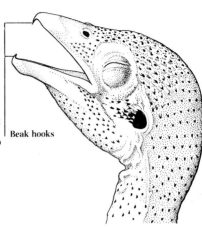

Beak hooks

GUIDING BEHAVIOUR

A honeyguide, having located a bees' nest, finds a ratel and attracts its attention by calling and tail fanning. It leads the ratel to the bees and waits nearby while the ratel breaks into the nest and reveals the food supply. This indicating behaviour seems to be largely instinctive on the part of the bee-eater with little learnt component.

Honeyguide

Ratel

The Duck-billed Platypus

Most mammals are placentals, that is, they develop in the mother's uterus attached to a placenta. Born relatively well-developed, they are suckled on mammary glands. There are two other types of mammals of a more primitive stock: the marsupials, such as the kangaroo and the koala, which suckle their young in a pouch, and the monotremes, of which there are only two types—the spiny anteaters, or echidnas, and the duck-billed platypus, *Ornithorhynchus anatinus*.

The monotremes have highly specialized adaptations for their particular ways of life, but are judged to be primitive in retaining many features normally considered characteristic of reptiles. The most startling of these is that they lay eggs, even though they have hair on their bodies and primitive mammary glands.

The female platypus usually lays two white leathery eggs at a time. After a week to ten days' incubation, a naked, blind youngster emerges from each egg. For the next four months the young lick milk that issues from slits in the mother's abdominal wall.

Since the discovery of the platypus approximately 200 years ago, scientists have been puzzled by its unusual combination of characteristics, particularly its head structure and feeding technique. While swimming with the aid of its webbed feet, the platypus uses its bizarre duck-like jaws to probe the river bed for its prey. With this large flattened beak, which is sensitive to touch, it picks up crayfish, worms and frogs along with a great deal of mud and sand. Grooves in the beak enable mud and sand to be filtered away, although some is swallowed with the prey.

The powerful flattened claws of the forepaws are specialized for burrowing in the banks of rivers. Short tunnels, so-called camping burrows, are made by both sexes. But only the females dig tunnels, up to 40 feet (12 metres) long with a nesting chamber at the end, which they line with wet grass and other vegetation. It is this wet nest material that stops the eggs from drying out before hatching.

There is only a single species of platypus, and it lives in warm lowland rivers as well as cool mountain streams throughout eastern Australia and in Tasmania.

When the first dried skin of a platypus arrived at the Natural History Museum in London in 1799, it was thought to be a fake—an animal resembling a beaver with a duck's bill stuck on to its body. Scissor marks still remain on that preserved skin, where investigators tried in vain to find hidden stitches.

The duck-billed platypus is about 2 ft (60 cm) long including the flattened, beaver like tail which is about 6 in (15 cm) long. It weighs about 4 to 5 lb (1.8 to 2.2 kg). Most of the structural modifications of the platypus are concerned with its semi-aquatic existence. It has short legs and feet with strong claws and efficient webbing: the claws are used for digging and the webs when swimming. A hunting dive lasts about 5 minutes and each swimming stroke ends with an upward movement of the legs to keep the buoyant platypus from surfacing. Its feeding technique on the river-bed is dependent on the soft, leathery duck-billed snout supported by the flattened, diverging jaw bones. There are nostrils at the upper end of the bill. Smell and taste are the most important senses when searching for food; while under water, the tiny eyes are covered by fur.

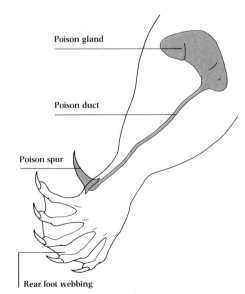

The hairy tail of the platypus seems of little use as a propulsive organ in water. It is thought that the tail is used to make an alarm signal slap on the water surface before diving to escape a potential attacker.

Poison gland

Poison duct

Poison spur

Rear foot webbing

A potent, curved poison spur points back from the ankle region of each hind leg of the male platypus. Each spur is connected to a poison gland in the soft tissues of the leg. If attacked, or in competition with another male, the platypus kicks back with both legs, lacerating its enemy and injecting poison. The poison can be harmful even to man but is never used in the capture of prey. The female platypus does not have this apparatus. Male spiny anteaters have similar spurs.

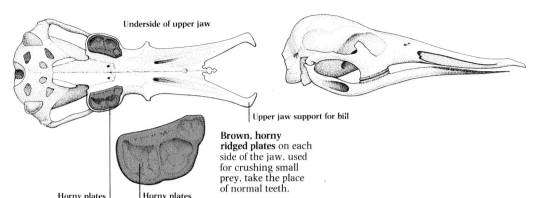

Underside of upper jaw

Upper jaw support for bill

Brown, horny ridged plates on each side of the jaw, used for crushing small prey, take the place of normal teeth.

Horny plates

Horny plates

The skull's most important feature is the adaptation of the jaw structure for the support of the duck-billed snout. The splayed premaxillae and dentary bones are not joined at the front and the soft, sensitive front of the bill is supported across this space.

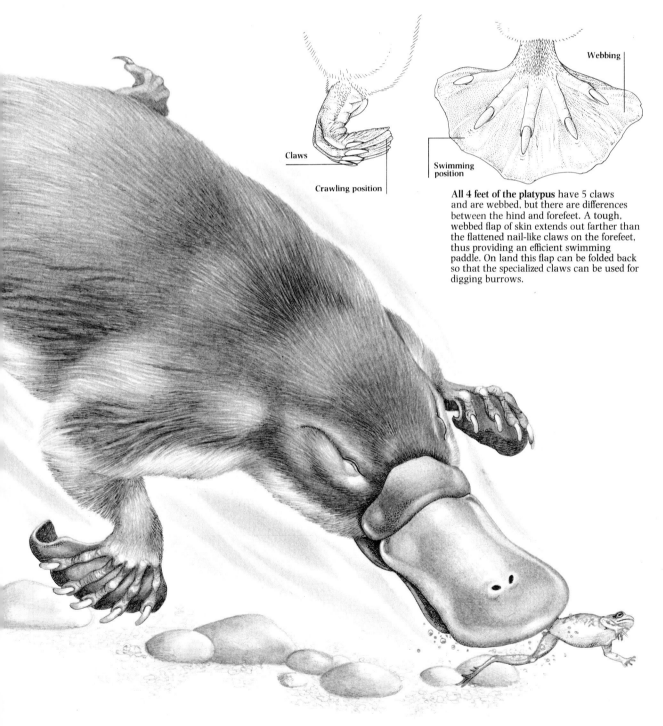

Webbing

Claws

Crawling position

Swimming position

All 4 feet of the platypus have 5 claws and are webbed, but there are differences between the hind and forefeet. A tough, webbed flap of skin extends out farther than the flattened nail-like claws on the forefeet, thus providing an efficient swimming paddle. On land this flap can be folded back so that the specialized claws can be used for digging burrows.

OPPORTUNIST HUNTERS: THE ALL-ROUNDERS

According to the rules of nature, predators must be opportunists if they are to survive. They must be able to adapt to changing climates, altering patterns of competition and drastic upheavals in the abundance of their common prey. And in today's world they must also be able to adjust to the new biological pressures caused by man. All—or nearly all—creatures that consume animal food must thus be able, if the occasion demands, to change their feeding habits, habitats or prey species. Animals like the Everglade kite and aardvark, that commit themselves to hunting a single prey type, are rare and well-known exceptions to the rule.

Some predators are extremely adept and sophisticated in paying the premiums of adaptability, the vital zoological insurance policy. These are the world's generalists, the opportunists which take advantage of almost any conceivable feeding situation or food supply. The only strategy that makes this responsiveness possible is the paradoxical one of lack of specialization. The successful generalist predators—animals like bears and foxes—refuse to commit themselves biologically. In structure, behaviour and physiology they keep their options open. The code that constrains their lives is that any extreme specialization which allows for one supremely efficient method of predation will diminish the ability to kill in other ways—and these other methods might, one day, mean the difference between extinction and survival.

Among the opportunists, omnivorous nutrition is common. Animal food is taken, but essential foodstuffs are also obtained from plants. Fruits of all kinds, which are rich in readily digested nutrients, are searched for regularly by these experts in adaptability. Equally, the sensory and offensive equipment of these predators is not highly specialized. Hearing, smell, taste and vision are often all moderately sensitive and no one of these senses is vastly more efficient than the others. Attacking weapons are also designed so that many sorts of prey can be taken in a variety of ways.

In the context of modern ecology, these qualitative descriptions of the life-styles of opportunists can be expressed more quantitatively. The totality of the way in which an animal makes its living is described as its niche, which includes where it lives, how and on what it feeds. By splitting up the analysis of an animal's niche into several resource gradients, each comprising some aspects of the environment for which the animal competes, such descriptions are made quantitative.

For a series of insectivorous birds, the size of insects is, for example, one resource gradient. Any bird in the series will normally eat insects of a size that falls within a characteristic range. The range describes just one resource—food—of the total niche. When animals compete fiercely within an ecosystem they tend to minimize the overlap in their use of resources. It is extremely unusual, for instance, for two different species of insect-eating birds, living in the same woods, to feed in the same parts of the same trees and to take insects of the same size. Thus, in a wooded area, blue tits will tend to eat smaller insects than great tits (two species of *Parus* or chickadees). And in these terms, the great asset of generalist predators is that they can use an exceptionally wide range of resources in their niches.

The Bear

No one could mistake a bear for any other kind of mammal. All bears are large, muscular, heavily built creatures, usually with shaggy coats. Although often regarded as fierce predators, bears are in fact utterly uncommitted generalists which, despite their fearsome teeth and claws, feed largely on fruit, nuts, underground plant tubers and insects. They will eat fish and carrion but only rarely attack, kill and eat prey of an appreciable size.

Anatomically, a number of structural features mark out the bears from other large predators. They walk on the flat of their feet and all four feet have long claws that cannot be retracted. Teeth are suited both for attacking prey and crushing vegetable matter. The tail is always very reduced in length and the eyes and ears are small. Bears are generally well adapted for life as unfussy foragers in wild upland country or forests.

The bears belong largely to the northern hemisphere. A few are found in India, Southeast Asia and South America but no natural bear populations inhabit Africa. Seven bear species are recognized today, of which the most widespread is the brown bear, *Ursus arctos*, found from Sweden across the USSR, and in Canada and North America. Small, declining groups hang on in mountainous areas in the Pyrenees and Italy and the brown bear also inhabits warmer climates in Syria, Pakistan and northern India.

The wide range of geographical and ecological situations in which the brown bear survives is tribute to an opportunist hunter of great adaptability. Within this range of habitats, the smallest subspecies are found in the hottest regions, the largest in the cold north. The Syrian form often weighs only 150 lb (68 kg) while the Kodiak brown bear of Canada tips the scales at 1650 lb (747 kg). These giants among brown bears, the largest land carnivores, can be 9 feet (2.7 metres) from head to tail.

Salmon is the staple food of many Alaskan brown bears in the summer—the home ranges of these bears correspond closely to the salmon spawning grounds. The bear flings the fish on to the river bank with a sweep of a forepaw or pins it down in the shallows with its jaws or paw.

Canine tooth   Molar tooth

The skull is a solid framework for the bear's massive head and jaws. The teeth are illustrative of the animal's varied lifestyle. Strong conical canines form attacking weapons but the cheek teeth are more suited for grinding food. The rear molars are certainly specifically used for crushing nuts and berries.

Claw

Sets of long-clawed digits terminate both hind and forelegs. These non-retractile claws—5 to 6 in (12.5 to 15 cm) long—are powerful weapons for gripping large prey, breaking into bees' nests or fishing. Young bears use the claws to climb trees.

Plant food may supply up to 70 per cent of a brown bear's diet but an immense range of animal prey is also eaten: mice, beetles, voles, ground squirrels, young boar, deer, elk and carrion.

The Stoat

Minks and martens, stoats and skunks, weasels, otters and badgers are all carnivorous mammals belonging to the family Mustelidae. The predators of this extremely successful family, many of which have lithe, slender bodies, range from the specialist to the generalist. The otter, *Lutra lutra*, for example, is a specialist that has become almost completely committed to the underwater hunting of fish. The bulk of an otter's diet is made up of fish, with animals like crayfish and frogs making up the balance. At the other end of the scale, the European badger, *Meles meles*, is a generalist which, like the brown bear, feeds more as a vegetarian than a carnivore.

Between these two extremes is a wide selection of other mustelids, most of which are predators able to capture a number of different types of prey. Among the best known of these intermediate generalists is the stoat or

ermine, *Mustela erminea*. Size for size, the elongated, graceful stoats are as potent as the big cats. Highly skilled killers, they can subdue and devour prey, like rabbits and hares, that are much bigger than themselves. They also hunt and eat many rodents, including voles, as well as birds such as partridges, pigeons and passerines.

Geographically, the stoat is a natural inhabitant of much of the northern hemisphere from Alaska through North America to Europe, Scandinavia, the USSR and Japan. It also has a toehold on the African continent in Algeria.

An adult stoat is about 17 inches (43 centimetres) long, including a tail of 4 to 5 inches (10 to 12.5 centimetres). The fur is reddish-brown with white underparts and the tail has a black tip. In the northerly, colder parts of its range, the stoat's coat changes in winter from red-brown to pure white (except for the black tail tip). This is the ermine fur

sought after and prized by the fur trade.

Stoats hunt by day or night. Searches for prey take place in many types of terrain—like all the best generalists the stoat is very adaptable and can make use of any habitat where prey may be present. In Europe and North America stoats do their hunting in fields and other grassland, along hedgerows, in scrubby vegetation or by rivers. The slim, supple, short-legged stoat can, in fact, move effortlessly through most sorts of cover. A stoat's quarry is found largely by means of the predator's keen sense of smell. The final, rarely unsuccessful attack consists of a rapid pounce terminated with an accurate death bite to the victim's throat or neck.

The European weasel, *Mustela nivalis*, and the two North American species, the least weasel, *Mustela rixosa*, and the long-tailed weasel, *Mustela frenata*, are mustelids with appearance and biology extremely similar to the stoat's.

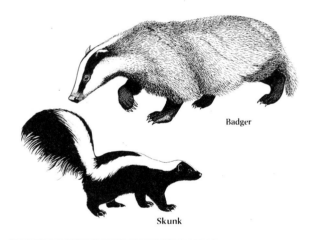

Badger

Skunk

Badgers are stocky mustelids with boldly striped heads. As well as plant food they eat worms, beetles, slugs, snails, bee grubs, rodents and hedgehogs. Skunks are also omnivorous. They are notorious for their defence mechanism—a nauseating smell produced from secretions in their anal sacs and squirted at an attacker.

COLOUR CHANGE
In both the stoat in Europe and the long-tailed weasel in the United States, a winter colour change from brown fur to white fur occurs only in the northern parts of the animal's range and at high altitudes. The normal coat is moulted and replaced by a new white one in as short a time as a month. The moult is under complex hormonal control, but surprisingly, the vital external signal for the coat change is the shortening of the days rather than the temperature drop.

Summer coat

Winter coat

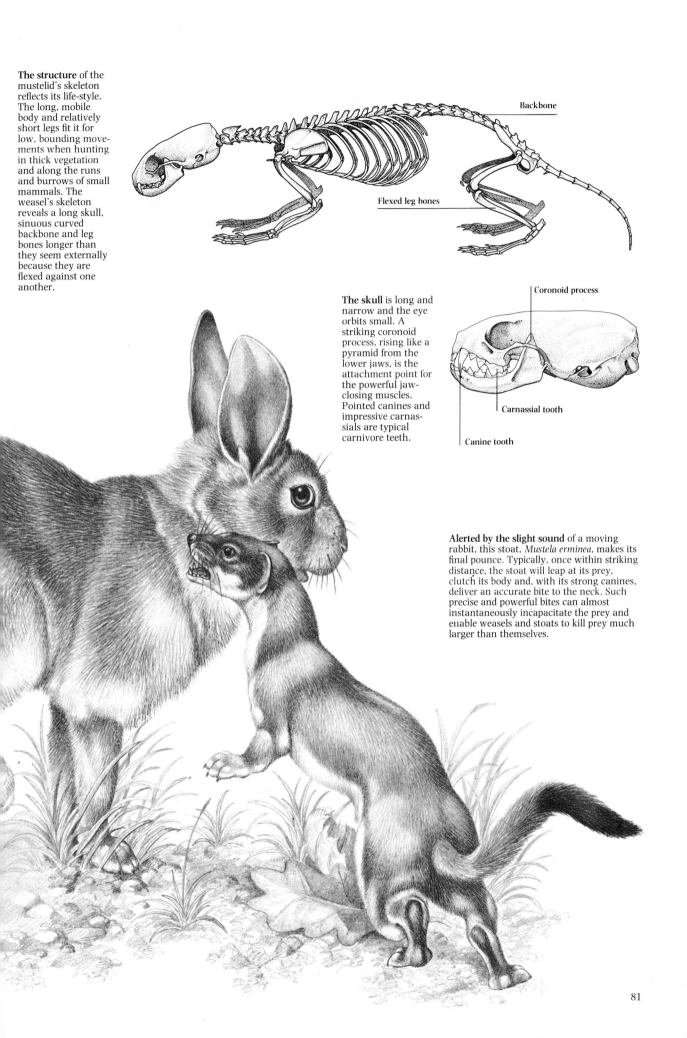

The structure of the mustelid's skeleton reflects its life-style. The long, mobile body and relatively short legs fit it for low, bounding movements when hunting in thick vegetation and along the runs and burrows of small mammals. The weasel's skeleton reveals a long skull, sinuous curved backbone and leg bones longer than they seem externally because they are flexed against one another.

Backbone

Flexed leg bones

The skull is long and narrow and the eye orbits small. A striking coronoid process, rising like a pyramid from the lower jaws, is the attachment point for the powerful jaw-closing muscles. Pointed canines and impressive carnassials are typical carnivore teeth.

Coronoid process

Carnassial tooth

Canine tooth

Alerted by the slight sound of a moving rabbit, this stoat, *Mustela erminea*, makes its final pounce. Typically, once within striking distance, the stoat will leap at its prey, clutch its body and, with its strong canines, deliver an accurate bite to the neck. Such precise and powerful bites can almost instantaneously incapacitate the prey and enable weasels and stoats to kill prey much larger than themselves.

OPPORTUNIST HUNTERS
The Baboon

Man and his relatives, including lemurs, bush babies, tarsiers, marmosets, monkeys and apes, are all primates—the "first ones"—occupying pride of place on the pyramid of animal evolution. Most primates are animals adapted for life in tropical forests and many species are most at home in the leafy canopies high above the forest floor.

The bones and muscles of the bodies of the arboreal primates are arranged to allow for a swinging, leaping and grasping existence among the tree tops. Food for these monkeys consists mainly of nuts, soft fruit and young leaves, shoots and branches of the trees in which they live. Forward-pointing eyes endow all primates with efficient binocular vision and this faculty in tree-dwellers, combined with hands ideally designed for precision grasping, makes item-by-item food foraging a most skilful, dextrous business.

Of the primates that have taken up

residence on the ground, man is the number one example. It seems highly probable that a crucial stage in the evolutionary development of this most complex primate was the abandonment of life in the trees. When man took to open country and became more predatory in his acquisition of food, the foundations of the basic traits in his sociological make-up were laid.

Man is not the only primate to have "come down" from the trees. The baboons (*Papio* spp), patas monkeys (*Erythrocebus patas*) and mandrills (*Mandrillus* spp) have also deserted the forests and become adapted for an infinitely more generalist way of life at ground level.

The baboons are Old World monkeys found in Africa and Asia where they live mostly in rocky uplands or savanna, and are among the biggest of the monkeys. Baboons sleep by night and search by day for their omnivorous diet

of fruit, roots, reptiles, insects and scorpions, which is irregularly expanded to include the nestlings and eggs of birds, and mammals. Most baboons live in large troops where they show complex social behaviour. Each troop has a "pecking order", or dominance hierarchy, headed by the oldest males. The baboon skull is rather dog-like with a long face, and all four legs are of equal length—an adaptation for quadripedal walking on the ground. The average male baboon weighs about 77 pounds (33 kilogrammes) and lives for 25 years or more.

One of the best-studied baboons is *Papio hamadryas*, the Hamadryas, or sacred baboon, so called because it was revered by the ancient Egyptians as the god Thoth. It is now found in rocky ravines of Ethiopia, Sudan and the Arabian sub-continent. Other baboons include the yellow baboon of West Africa and the chacma baboon.

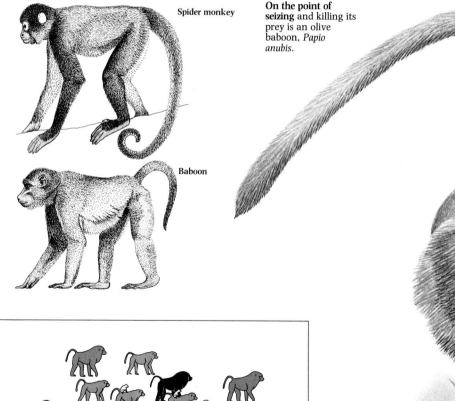

The tree-living spider monkeys brachiate, that is, swing, with alternate left and right hand holds, between branches. These monkeys have extremely long limbs, small thumbs (they hook fingers over the branches they swing on), and the structure of the chest wall and pectoral girdle allows the forelimbs to reach in all directions. Ground-living monkeys like baboons run on straight legs. Their rib cage and shoulder blades are designed for limb movement beneath the body and the thumbs are long for precision gripping.

Spider monkey

Baboon

On the point of seizing and killing its prey is an olive baboon, *Papio anubis*.

A BABOON TROOP

Baboons are social animals and live and hunt in large troops. One or two old males, which have subordinated all the other males in fights, dominate the group. When moving around, the troop is arranged with subordinate males at the front and rear of the group and dominant males, females and young at the centre. If attacked, all the males move toward the aggressor.

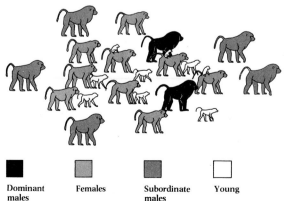

Dominant males

Females

Subordinate males

Young

The head of the baboon, and of other mandrill-like monkeys which have come out of the trees to live on open ground, is unlike that of most other Old World monkeys. Instead of the flat, short-snouted face of the latter, baboons have long jaws and an almost dog-like appearance. This structure relates to the development of long rows of molar teeth for grinding food. The upper canine teeth are large and pointed, making good weapons. Baboons have retained the forward-facing eyes typical of primates and the shape of the head facilitates uninterrupted vision.

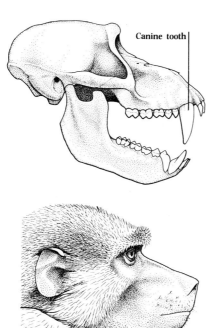

Canine tooth

Primates' specialized hands are important both for grasping branches and for handling food. Some, such as marmosets, have digits all functioning in the same way: they grasp objects between the palm and the 5 fingers. Other American monkeys (Super-family Ceboidea) have a thumb which they can separate from the other digits and bend independently. Baboons and other Old World monkeys and apes have the most complex hands, with a thumb which can be rotated about its base and placed flat against a finger to pick up small food items.

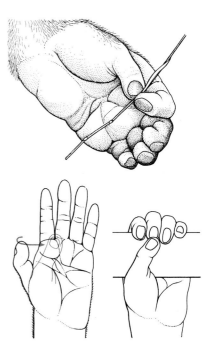

The Red Fox

The cunning and intelligence of the fox are proverbial. Although it is hard to conceive of an animal being cunning in the human sense, the red fox is, without doubt, an exceedingly adaptable and successful generalist predator.

The red fox has passed the most severe test that can be imposed on any carnivore—the sudden disappearance of one of its major prey species. When the virus disease myxomatosis was introduced into Britain in the early 1950s, rabbits essentially vanished from the countryside. Until this time, rabbits had been one of the staple foods of the fox, but the decimation of the rabbit population caused no great concern to this opportunist hunter. In some areas the shortfall was made up with extra vole kills and attacks on birds; in others, insects suddenly came to compose an appreciable part of the red fox's diet. On the Swedish island of Götland, although myxomatosis reduced the rabbit population by 95 per cent, the resourceful foxes still made rabbit kills at 50 per cent of the pre-myxomatosis level by devoting more attention and skill to catching the few remaining rabbits.

Extensive studies of the red fox and its food in many parts of the world, including natural fox populations in Britain, Sweden and the United States and introduced ones in Australia, have shown that this predator concentrates mainly on small rodents, rabbits and hares. But a picture also emerges of a versatile hunter that will eat, if the opportunity presents itself, carrion, ground nesting birds and their young, birds' eggs, fruit, sheep placentas in the lambing season, beetles and, increasingly in cities, the contents of garbage cans.

The fox does not, however, feed indiscriminately, but has overriding preferences that make it seek out particular types of prey, even if these are not very common. Maybe the fox is unusually efficient at killing these favoured prey species, or perhaps they simply taste better! With a range of small vole and mouse species to choose from, for example, the red fox will take significantly more voles of the genus *Microtus*, even when other voles like *Clethrionomys* and mice such as *Apodemus* are more abundant.

The red fox, *Vulpes vulpes*, is about 3 feet (90 centimetres) long, including a tail of about a foot (30 centimetres). The underside of its reddish-brown coat is whiter in colour, the backs of the ears black. The tail has a white tip. Throughout the northern hemisphere—in both cold and temperate regions—as far south as Africa and Southeast Asia, it lives in burrows.

The adaptable fox has perfected different food-handling techniques in order to take best advantage of what food is available. It can, for example, eat eggs with great delicacy and precision. Squatting on its hind legs, the fox holds the egg in its front paws, cracks the top of the shell with its front teeth and laps out the contents. Food burying is a common habit. Foxes bury surplus food, often eggs and carrion, under a shallow layer of soil to be recovered when they need it.

THE "MOUSE JUMP"

Mice and voles move in narrow runways in the vegetation where it is hard for the fox to follow. So, the fox locates a mouse by sight and sound, then leaps high into the air to land right on its victim, with paws and muzzle contacting the ground (and prey) at once.

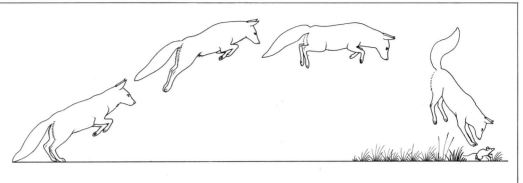

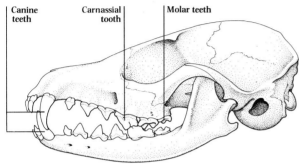

| Canine teeth | Carnassial tooth | Molar teeth |

The skull, particularly the teeth, of the fox is illustrative of its opportunist life-style. The canine teeth are a moderate size, but not highly specialized like, for example, those of the big cats; they are good, all-purpose attack weapons. The molar teeth are best suited for crushing vegetable food.

Kit fox

Bat-eared fox

The kit fox, also known as the swift fox, *Vulpes velox*, lives in the Great Plains area of North America and feeds on rabbits, hares, rodents and birds. The bat-eared fox, *Otocyon megalotis*, of southern and eastern Africa, is the most insectivorous member of the dog family. With its huge ears, it locates insects under the ground by the sounds they make, particularly termites and beetle larvae.

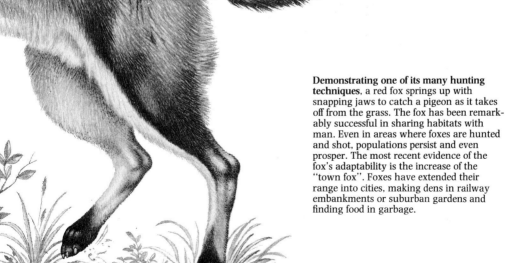

Demonstrating one of its many hunting techniques, a red fox springs up with snapping jaws to catch a pigeon as it takes off from the grass. The fox has been remarkably successful in sharing habitats with man. Even in areas where foxes are hunted and shot, populations persist and even prosper. The most recent evidence of the fox's adaptability is the increase of the "town fox". Foxes have extended their range into cities, making dens in railway embankments or suburban gardens and finding food in garbage.

The white-tipped tail of the red fox is believed to be used for signalling. Near the base of the tail are openings from anal sacs which contain a strong-smelling secretion. The fox lifts its tail and rubs this area against trees or fences to mark out its territory.

The Carrion Crow

Of all the many groups of birds that populate the earth and its skies the crow family, Corvidae, is undoubtedly among the most advanced and successful. The accomplishments of the family can be judged by the plethora of species—103 are found worldwide—and the way in which they have spread to almost every corner of the earth.

Despite the human associations, it is difficult not to apply the word intelligent to crows, the birds which, it is suggested, represent today's pinnacle of avian evolution. The versatile repertoire of communal life and feeding behaviour they employ in their adaptable life-styles suggests that much of their behaviour is learnt rather than instinctive. Indeed, some species demonstrate social organizations of a complexity unknown in the rest of the bird world.

This glowing description of the successful attributes of the crows might suggest that they are highly specialized, but this is not so. They are, in fact, extremely adaptable generalists in both

structure and behaviour, and by being so are able to keep most of their biological options open. Nearly all crows are omnivores, feeding on both plant and animal foods.

Compared with their fellow members of the order Passeriformes—the huge and important group of perching birds—crows are generally large in size. The raven, which measures just over 2 feet (60 centimetres), is the largest perching bird in the world. In almost all crows the beak is moderately long, straight and powerful—a general-purpose foraging and attacking tool. The nostril openings at the base of the beak are protected by forward-facing, stiff bristles. Ravens, rooks, jackdaws and crows have more or less uniformly black plumage, but other, less advanced members of the family, such as jays, are more colourful and strongly marked.

The American and Old World jays, the most colourful of all the crows, include the blue jay, *Cyanocitta cristata*, of North America and the common jay,

Garrulus glandarius, which is widespread throughout Europe and Asia. A woodland bird, the common jay feeds on acorns for much of the year, but in the breeding season becomes far more rapacious and steals both the eggs and young of other birds from their nests.

The magpies are crows with strongly marked plumage and long, straight tails. A stock example is the magpie, *Pica pica*, found all over the north of the Old World and the western parts of North America. Like the jays, magpies have strongly predatory tendencies, but, unlike them, are ground feeders.

A typical representative of the true black crows, probably the most advanced members of the crow family, is the carrion crow. This subspecies of *Corvus corone* breeds in England and Wales, France, Germany, Spain and Portugal. Often supplementing their food intake by scavenging from man's garbage dumps, carrion crows are foragers and predators with wide-ranging omnivorous diets.

In its bone pattern, a bird's skull is similar to those of its ancestors, the reptiles. The skull is light, because of load problems, and has a compact rounded braincase. The most prominent features are the beak bones which in life are sheathed with a horny substance.

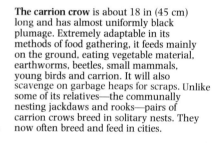

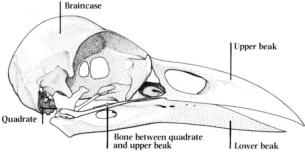

The beak, and its connections with the rest of the head, form a complex system of hinges and levers with a considerable degree of movement between the bones. The mobile framework contrasts with the rigid skulls of reptiles. This system allows the upper bill to move during beak opening as well as the lower bill. As the beak opens the lower jaw drops and pivoting of the jaw around a ligament fulcrum rocks an attached bone, the quadrate. This in turn pushes a bony rod. The rod is hinged to the back of the upper half of the beak, and as it moves forward it makes the upper jaw swing upward, thus increasing the gape and the capacity to swallow large prey.

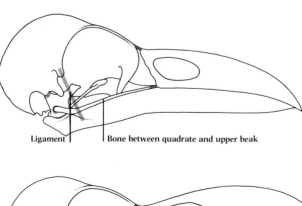
Braincase / Upper beak / Quadrate / Bone between quadrate and upper beak / Lower beak

Ligament / Bone between quadrate and upper beak

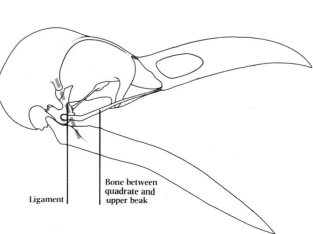
Ligament / Bone between quadrate and upper beak

The carrion crow is about 18 in (45 cm) long and has almost uniformly black plumage. Extremely adaptable in its methods of food gathering, it feeds mainly on the ground, eating vegetable material, earthworms, beetles, small mammals, young birds and carrion. It will also scavenge on garbage heaps for scraps. Unlike some of its relatives—the communally nesting jackdaws and rooks—pairs of carrion crows breed in solitary nests. They now often breed and feed in cities.

Reduced tail

Femur

Tibia

Sternum keel for wing muscles

Fused metatarsals

The skeleton of the crow serves as a generalized diagram of the structure of a basic bird, so unspecialized is its body architecture. It demonstrates how the forelimb is radically adapted to provide a framework for the wing feathers, and how the hind limb is in 3 sections: femur, tibia and fused metatarsals. Also striking is the huge enlargement of the breastbone (sternum) keel, as an attachment point for the large pectoral muscles which power the beating of the wings.

Hooded crow

Although their plumage is different, the hooded crow and the carrion crow are geographically separate subspecies of the same species, *Corvus corone*. The hooded crows are found in Ireland, North Scotland, Scandinavia, Eastern Europe and Russia.

Despite its name, the carrion crow is a rapacious killer of live prey, particularly young birds. No eggs or nestlings of ground-nesting birds are safe from this predator.

SPECIAL SENSES: BIZARRE MECHANISMS

The Rattlesnake

At the heart of every attempted act of predation lie two technical problems—finding the prey and killing it. In solving these biological questions, predatory animals, and the prey creatures they habitually kill, have had massive evolutionary effects on one another. In the process of evolution the predators and prey most likely to survive and be parents of future generations are those with genetic constitutions that enhance, respectively, hunting efficiency and the evasion of capture.

The conflicting evolutionary pressures benefiting both predators and prey have helped shape the near-miraculous prey-finding machinery of some predators. This machinery consists of an integration of hunting behaviour with sensory organs, the structures connected to an animal's central nervous system which signal subtle or gross changes in its surroundings. Sense organs, or receptors, are transducers of energy, each specialized to receive information about a particular type of change in the environment and convert it into signals that the body can understand. The changes may be in temperature, light, colour, touch, chemicals or sound, but in the receptor organ these are converted into coded messages in the form of nerve impulses. These patterned arrays of impulses, the vocabulary and grammar of the nervous system, are sent to the brain where they are interpreted as the languages of vision, hearing, smell, taste and so on.

In the vertebrates, refinements of the sense organs for hunting either increase the absolute sensitivity of the receptors or make these animals responsive to environmental clues that are useless or meaningless to most other creatures. Both types of adaptation are common in specialist carnivores. Eyes, ears, nose and organs of taste and touch may become bigger or more complex in internal construction, but completely original sensory systems may also be developed.

While predators hunt in normal conditions, standard hunting behaviour and sense organs suffice for successful predation. But in unusual and difficult sensory circumstances, new or ultrasensitive receptors become crucial to prey finding. Many of the most bizarre examples of predator receptor organs are found in hunters that try to locate their prey in testing surroundings. How, for instance, can a predator find its prey in complete darkness where it is denied the use of vision? A few specialist carnivores have evaded the difficulty by producing their own light. Several deep-sea fish, which hunt in the lightless ocean depths, have light-emitting organs which, at close range, illuminate prey like diffuse search beams. These answers to sensory problems are mirrored by some invertebrates: deep-sea squids may have light-producing organs like those of abyssal fish.

Other predators use alternative sensory systems that open up areas of environmental knowledge denied to their rivals. Owls can use redesigned ears that give accurate locations of sound sources in complete blackness. Bats bounce short-wavelength (ultrasonic) sound pulses off flying prey in a type of airborne radar that is efficient in the blackest night. In an unsurpassed way, some snakes such as the rattlesnake have evolved "heat camera" eyes that can pinpoint the heat (infrared rays) emitted by warm-blooded prey.

The snakes have always had a special place in human consciousness as death-dealing symbols of evil. Certainly the virulent venoms of quite small snakes can quickly kill a man, but not all snakes are venomous. Snakes are essentially of two types. The relatively primitive group that includes the boas and pythons have neither poison nor poison-injecting fangs. Instead they crush and chew their prey to kill it. The cobras, vipers and other more advanced snakes are the ones that have become highly specialized for killing prey with venom. In these forms much of the anatomy of the head is modified for the swift, accurate delivery of venom into the prey during an attack.

Of all snakes, the vipers are the most modified. Among them are the true vipers of the Old World, such as the adder, *Vipera berus*, and the rattlesnakes, *Crotalus* spp, of the Americas. Rattlesnakes hunt their vertebrate prey by day and night. Toads, frogs, birds and small mammals such as mice are all assaulted with lightning-swift strikes and rapidly killed by the venom injected from a pair of tubular, hypodermic fangs set in the bones of the upper jaw.

Rattlesnakes possess two novel systems of information gathering that help them locate their prey in difficult conditions, as at night. They have "eyes", or pits, sensitive to infrared radiation that enable them to "see" patterns of the heat emitted in the world around them. They also have a special technique for picking up the track of a prey. The forked tongue, itself an organ of taste, is flicked out to pick up chemical clues which it deposits in two small cavities called Jacobson's organ in the roof of the mouth. The organ is thought to be able to identify these clues as the particular "smell-taste" of regularly hunted prey.

Many small animals fall victim to the rattlesnake. At night the snake's infrared "eyes" can discern a warm-blooded prey animal, such as a bird or mouse, against its cooler background.

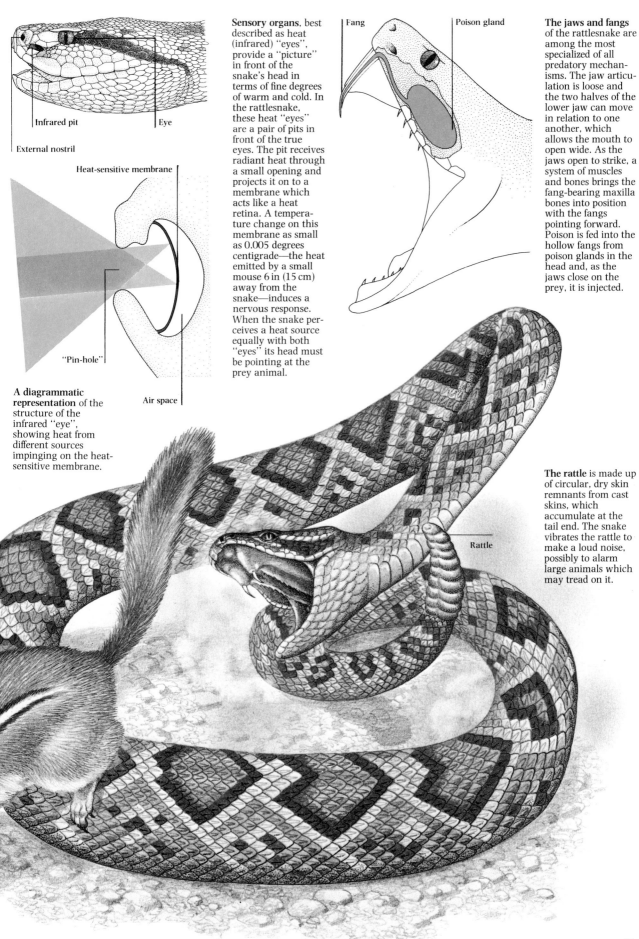

Sensory organs, best described as heat (infrared) "eyes", provide a "picture" in front of the snake's head in terms of fine degrees of warm and cold. In the rattlesnake, these heat "eyes" are a pair of pits in front of the true eyes. The pit receives radiant heat through a small opening and projects it on to a membrane which acts like a heat retina. A temperature change on this membrane as small as 0.005 degrees centigrade—the heat emitted by a small mouse 6 in (15 cm) away from the snake—induces a nervous response. When the snake perceives a heat source equally with both "eyes" its head must be pointing at the prey animal.

Infrared pit

Eye

External nostril

Heat-sensitive membrane

"Pin-hole"

Air space

A diagrammatic representation of the structure of the infrared "eye", showing heat from different sources impinging on the heat-sensitive membrane.

Fang

Poison gland

The jaws and fangs of the rattlesnake are among the most specialized of all predatory mechanisms. The jaw articulation is loose and the two halves of the lower jaw can move in relation to one another, which allows the mouth to open wide. As the jaws open to strike, a system of muscles and bones brings the fang-bearing maxilla bones into position with the fangs pointing forward. Poison is fed into the hollow fangs from poison glands in the head and, as the jaws close on the prey, it is injected.

Rattle

The rattle is made up of circular, dry skin remnants from cast skins, which accumulate at the tail end. The snake vibrates the rattle to make a loud noise, possibly to alarm large animals which may tread on it.

The Barn Owl

The owls are the avian overlords of the night air—the only birds, apart from the nightjars and their close relatives, which have achieved notable success as nocturnal predators. Nearly all other carnivorous birds are daytime hunters for whom accurate eyesight is crucial.

The nightjars, such as the European *Caprimulgus europaeus* and the American poor-will, *Phalaenoptilus nuttallii*, "trawl" for flying moths and other insects. They have huge eyes and take to the night sky with their mouths agape ready to engulf passing prey, but their activities pale in comparison to the nocturnal raiding prowess of the owls.

Even in complete darkness, the owls can successfully hunt and kill. This supreme efficiency is made possible by a wide range of adaptations including huge, ultrasensitive eyes which are responsive to even the lowest light levels. These eyes are housed in an immensely mobile head which can rotate through 180 degrees or more—effectively allowing the owl to look directly behind itself. In the complete absence of light, however, owls depend on their hearing. Their ears can pinpoint the sources of minute sounds with extreme precision. This ability enables an owl to home in and plant its claw-tipped talons on a mouse at the end of an attack swoop without using its excellent powers of vision at all.

Soft-edged feathers are another adaptation for nighttime predation. These feathers cut down the noise of flight and allow the owl to make an almost silent approach.

Owl species around the world have a variety of plumage. Many species are coloured for camouflage in streaky or barred browns, greys and black. The owls of the cold northern forests are grey while the black-flecked white feathers of the snowy owl, *Nyctea scandiaca*, help it to merge with its snow-covered habitat. Such camouflaging plumage gives protection during long, daytime waits spent in hiding among ground vegetation or in tree holes. Although some owls hunt in daylight most stay hidden until dusk.

Owls vary in size from the elf owl of North America, *Micrathene whitneyi*, which is 5 inches (12.5 centimetres) long, to the appropriately named eagle owls, like the European *Bubo bubo*, which is 28 inches (70 centimetres) long. While the elf owls nest in cacti and feed on insects, eagle owls will kill birds and mammals the size of a roe deer.

Among the most typical owls are the medium sized tawny owl, *Strix aluco*, and barn owl, *Tyto alba*, which measure 12 to 15 inches (30–38 centimetres). Except for a few powerful bats, these owls are the only nocturnal, airborne hunters to take advantage of small mammals as their chief source of food.

Vision is of paramount importance to the owl's hunting technique. The owl has tubular eyes which depart from the normal globular shape. The eyes are positioned in facial discs—zones of flattened feathers fringing the beak. They point forward and must provide stereoscopic vision.

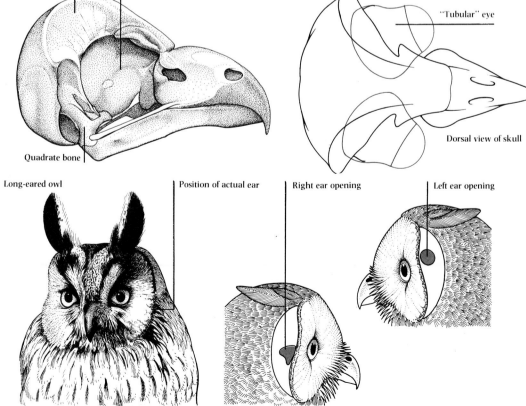

Braincase | Eye orbit area

Quadrate bone

"Tubular" eye

Dorsal view of skull

The "ear tufts" on the heads of some owls are actually nothing to do with hearing but are probably a form of display. The real ears are hidden under feathers at the sides of the skull. In some species the ears are asymmetrical.
This is thought to be linked to the owl's ability to pinpoint sound sources. By rocking its head, it may locate sound in the vertical as well as the horizontal plane.

Long-eared owl

Position of actual ear | Right ear opening | Left ear opening

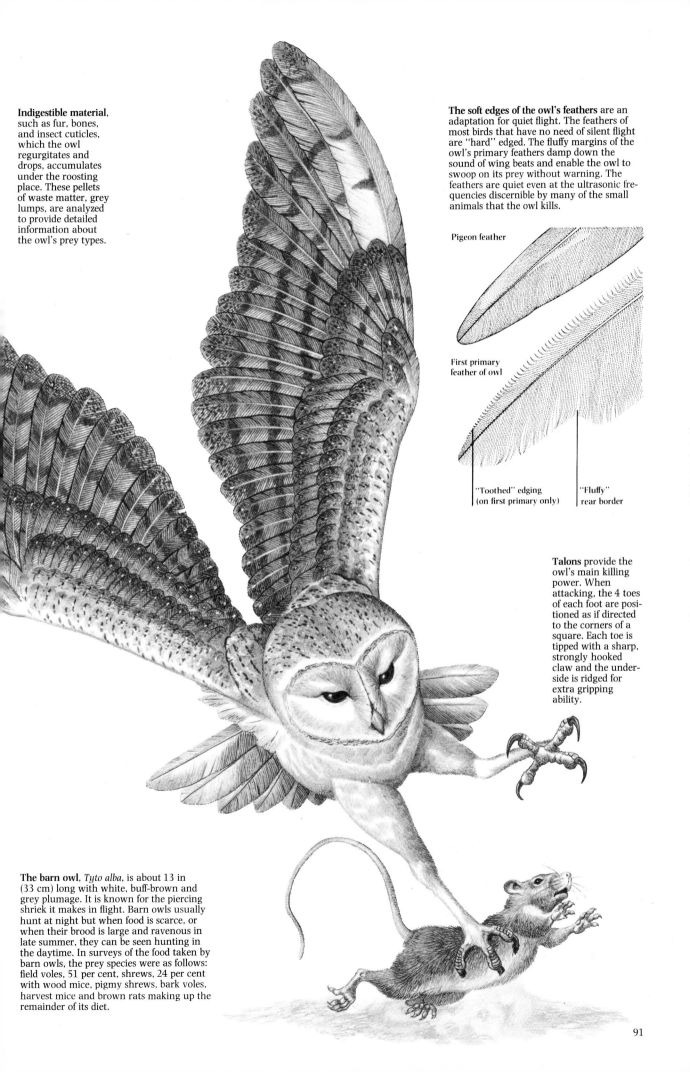

Indigestible material, such as fur, bones, and insect cuticles, which the owl regurgitates and drops, accumulates under the roosting place. These pellets of waste matter, grey lumps, are analyzed to provide detailed information about the owl's prey types.

The soft edges of the owl's feathers are an adaptation for quiet flight. The feathers of most birds that have no need of silent flight are "hard" edged. The fluffy margins of the owl's primary feathers damp down the sound of wing beats and enable the owl to swoop on its prey without warning. The feathers are quiet even at the ultrasonic frequencies discernible by many of the small animals that the owl kills.

Pigeon feather

First primary feather of owl

"Toothed" edging (on first primary only)

"Fluffy" rear border

Talons provide the owl's main killing power. When attacking, the 4 toes of each foot are positioned as if directed to the corners of a square. Each toe is tipped with a sharp, strongly hooked claw and the underside is ridged for extra gripping ability.

The barn owl, *Tyto alba*, is about 13 in (33 cm) long with white, buff-brown and grey plumage. It is known for the piercing shriek it makes in flight. Barn owls usually hunt at night but when food is scarce, or when their brood is large and ravenous in late summer, they can be seen hunting in the daytime. In surveys of the food taken by barn owls, the prey species were as follows: field voles, 51 per cent, shrews, 24 per cent with wood mice, pigmy shrews, bark voles, harvest mice and brown rats making up the remainder of its diet.

The Horseshoe Bat

Bats, the only flying mammals, are extraordinarily successful creatures. Their zoological group, the order Chiroptera, contains about 800 separate species spread around the world. The success of the bats is simple to explain. Because they can fly superbly—as well as birds—and also catch airborne food in complete darkness, they have virtually no competitors.

Insect-eating is the essence of existence for many bats, such as the horseshoe bat, *Rhinolophus ferrumequinum*, and seems to have been the earliest way of life in chiropteran evolution. Exactly when the bat experiment in winged mammalian life began is not known, but bat fossils 50 million years old look remarkably like the bats of today; so the first bat-like creatures must have split from a group of flightless insectivores at some point before this.

The fragile membrane of a bat's wing is stretched between modified, elongated limbs. These wings are not only adaptable organs of flight but also act as voluminous catching nets for flying insects. Yet the power of flight is only half the story of the bat's uniqueness, for nearly all bats possess a sophisticated form of radar, a mechanism of navigation and food location whose bioengineering bears an uncanny resemblance to man's radar systems.

Unlike man-made radar, the bat's system works with emitted pulses of sound, and in this sense is more analogous to underwater sonar. Pulses are sent out through the flying bat's mouth or nose into the airspace in front of it and bounce off obstructing objects or flying prey items such as moths. The returning echoes are picked up with large antenna-like ears and analyzed to give information about the size, position and relative speed of the object from which the pulses have been reflected.

Physical principles make important constraints on the type of sound that bats can use. For echo location the crucial quality of sound is its wavelength, the distance between two adjacent peaks on a sound wave. Sounds that man can hear have wavelengths too long to bounce effectively off small objects. If the wavelength is longer than the size of an object, the waves are not reflected but "ooze" round it. But ultrasonic sound, too high pitched for man to detect, has wavelengths short enough to bounce off objects the size of a moth. It is precisely these high pitched sounds which are used by bats.

Mediterranean horseshoe bat

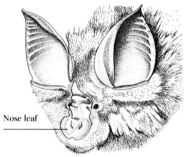

Nose leaf

Mouse-eared bat

A bewildering variety of bat species exists. The faces are particularly variable and these differences are connected with the ultrasound system. Some bats hum and snort their pulses through the nose; these species have complex nostrils forming a "nose leaf", which directs and modifies the sound pulse. Others shout their ultrasound through open mouths. In such forms the nose is simple and the snout resembles that of a normal insectivorous mammal. The external ears of bats are large and complex to pick up echoes returning from prey.

The fabric of the wings is a thin meshwork of muscle strips covered by skin. The wings are elastic and supple, enabling the bat to perform acrobatic feats in pursuit of its prey.

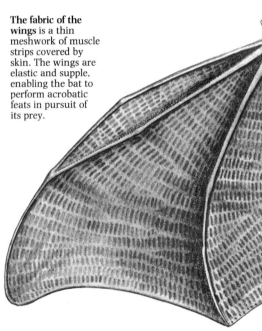

Extreme adaptations for flight are evident from the bat's skeleton. The shoulder girdle and forelimbs are huge and are the main structural members of the wings. The ulna bone is puny and the slim humerus and radius bones support the front of the wing skin. Long, out-stretched "fingers" carry the outer section of the wing and the thumb is reduced to a small hook at the front edge of the wing. The pelvic girdle and hind legs are small.

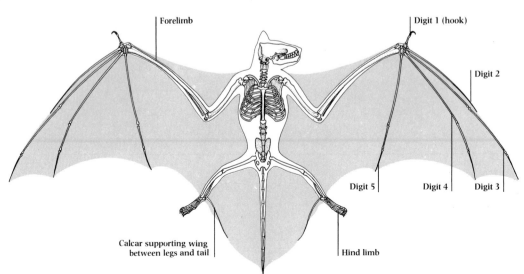

Forelimb

Digit 1 (hook)

Digit 2

Digit 5 — Digit 4 — Digit 3

Calcar supporting wing between legs and tail

Hind limb

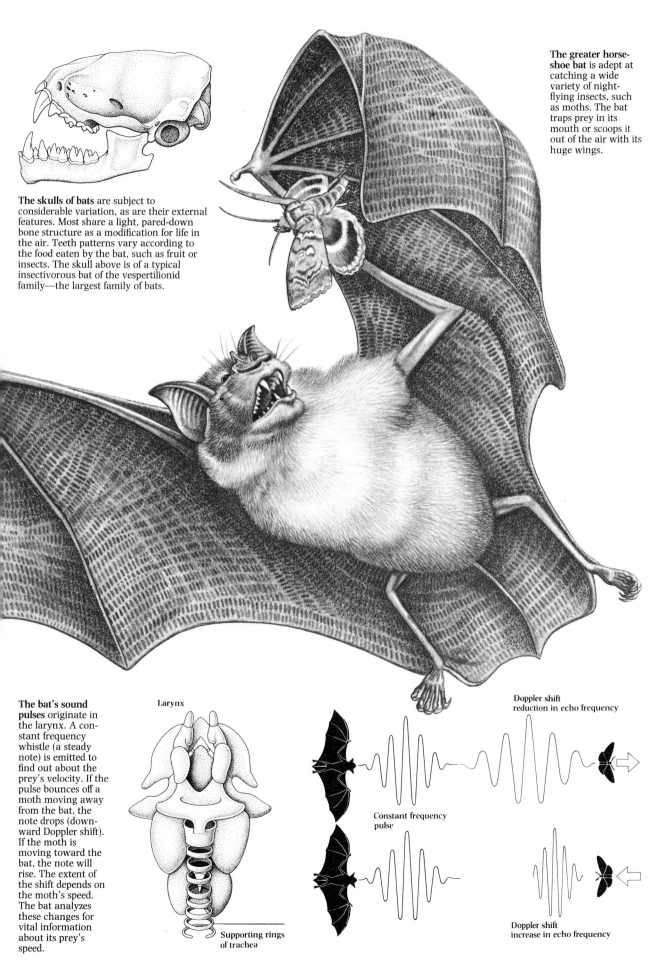

The skulls of bats are subject to considerable variation, as are their external features. Most share a light, pared-down bone structure as a modification for life in the air. Teeth patterns vary according to the food eaten by the bat, such as fruit or insects. The skull above is of a typical insectivorous bat of the vespertilionid family—the largest family of bats.

The greater horse-shoe bat is adept at catching a wide variety of night-flying insects, such as moths. The bat traps prey in its mouth or scoops it out of the air with its huge wings.

The bat's sound pulses originate in the larynx. A constant frequency whistle (a steady note) is emitted to find out about the prey's velocity. If the pulse bounces off a moth moving away from the bat, the note drops (downward Doppler shift). If the moth is moving toward the bat, the note will rise. The extent of the shift depends on the moth's speed. The bat analyzes these changes for vital information about its prey's speed.

Larynx

Supporting rings of trachea

Constant frequency pulse

Doppler shift reduction in echo frequency

Doppler shift increase in echo frequency

The Flea

It is hard to be dispassionate about fleas, insects whose very name can produce an itch. Fleas are certainly a dangerous, uncomfortable nuisance to man and the rat flea, carrier of the deadly bubonic plague, is more than a pest, it is a scourge. Yet fleas are supremely adapted animals whose bodies, behaviour and development are completely committed to an ectoparasitic life-style centred round bouts of blood-feeding on the skins of mammals and birds.

Because the diet of fleas is totally restricted to the blood of mammals and birds—the last two groups of vertebrate animals to appear on earth—fleas are probably a relatively recent evolutionary "invention" whose differentiation has developed in parallel with that of their warm-blooded hosts. One striking anatomical feature of a flea—its flattened body compressed from side to side—is an adaptation specifically designed for easy creeping between the feathers or hairs on the host surface.

The life of a flea begins when eggs are laid on the host's body. These eggs soon drop off to develop on the floor of the host's burrow, lair, nest or house. The minute, white maggots that hatch from the eggs feed directly from the host or on a variety of organic rubbish, including the droppings of adult fleas which contain host blood.

Usually about $\frac{1}{16}$ inch (1.5 millimetres) long, adult fleas are brown, flattened insects that never have wings. The external body covering, the cuticle, is remarkably tough—it is hard to crush a flea between finger and thumb. Using their powerful hind legs, fleas jump on to their hosts where the double-hooked tips on each of the flea's six legs give a secure grip on the skin. Piercing and sucking mouthparts then move into action for obtaining blood from vessels near the host's skin surface. The blood of a wide range of host species often suffices, but for some fleas only the blood of one type of host is suitable to ensure survival and reproduction.

Fleas can survive for long periods, perhaps over a year, without feeding on host blood. During this enforced starvation, the fleas wait patiently in the nest or burrow for the return of the host—and the blood it contains. When the host does reappear, the fleas are able to home in on it by using sensitive and sophisticated sensory abilities. In this recognition process, fleas can apparently smell their hosts, but detection of warmth and carbon dioxide are also crucial.

The bodies of both mammals and birds are maintained at a temperature of about 96 to 108 degrees Fahrenheit (36 to 42 degrees centigrade), which makes them, in most circumstances, hotter than their immediate surroundings. Fleas can locate these "hot spots" close by them, then jump towards their host. Similarly, respiring birds and mammals push carbon dioxide into the air around them from their lungs and skins. This contamination of the air, and the air currents created by a moving host can be sensed and used by fleas seeking a new food source.

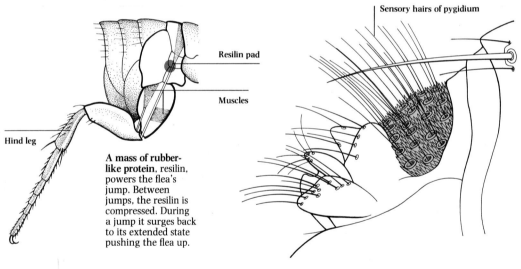

Sensory hairs of pygidium

Resilin pad

Muscles

Hind leg

A mass of rubber-like protein, resilin, powers the flea's jump. Between jumps, the resilin is compressed. During a jump it surges back to its extended state pushing the flea up.

At the rear of the abdomen, on the dorsal surface, is a complex of sensory organs used for finding a host. The most important is a group of cuticular hairs, the pygidium. These seem to be sensitive to air currents caused by the movement of the host animal, so that the flea knows when the host is near. The abdomen also contains parts of the gut which can be expanded to accommodate blood meals.

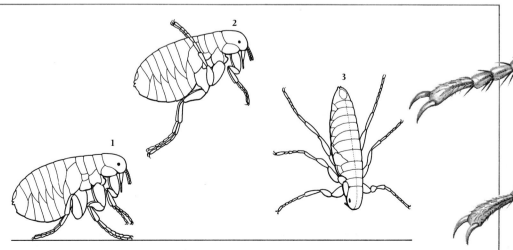

A FLEA JUMP

A flea can jump a distance 300 times its own length. The immense acceleration and power of the jump causes the flea to cartwheel through the air, turning end over end. To ensure a secure landing, the first or second pair of legs is held up above the body, with the tarsal hooks pointing forward. These act as anchors in the event of an upside-down landing.

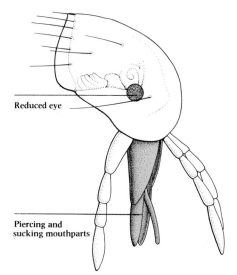

Reduced eye

Piercing and
sucking mouthparts

A flea's head has a range of adaptations for its blood-feeding existence. The 2 antennae are club-shaped and can be retracted, like an aircraft undercarriage, into grooves at the side of the head. This protects them when the flea is pushing its way through hair or feathers. On the underside of the head is a veritable dissection kit of mouthparts which probe and pierce the host's skin, then form a syringe for sucking up blood from the host.

Rows of evenly spaced, tooth-like spines are a modification for the flea's parasitic life-style. The cuticular "combs" act as anchors to lock the flea in among the fur or feathers of its host. Many fleas have a row on the head (the genal comb) and another at the front of the thorax (the pronotal comb). Research has shown a distinct connection between the spacing of the comb spines and the thickness of the host's hair. The finer the hair, the smaller are the gaps between the "teeth" of the comb.

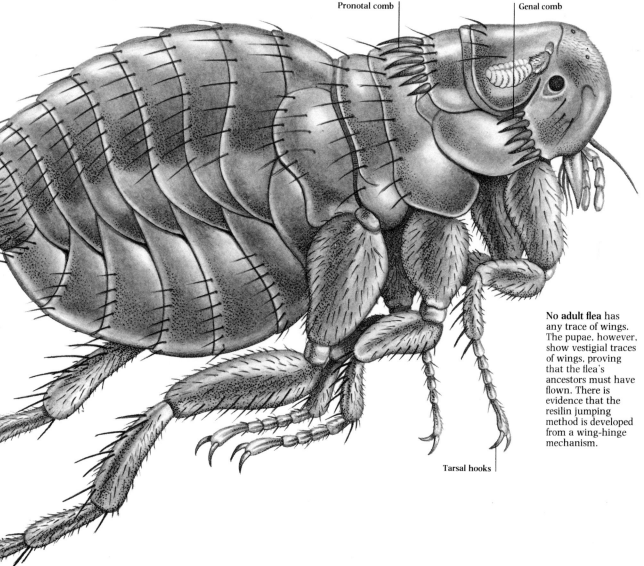

Pronotal comb

Genal comb

No adult flea has any trace of wings. The pupae, however, show vestigial traces of wings, proving that the flea's ancestors must have flown. There is evidence that the resilin jumping method is developed from a wing-hinge mechanism.

Tarsal hooks

The rabbit flea, *Spilopsyllus cuniculi*, is one of the most intensively studied species. It is about $\frac{1}{16}$ in (1.5 mm) long; the rabbit flea above is magnified 120 times. Among the extraordinary facts about the life-style of

this flea are those concerning the synchronization of its reproduction with the breeding of its host. The ovaries of the female flea do not develop until she is feeding on blood from a pregnant rabbit. Hormonal changes

in the rabbit's blood trigger off reproductive development and breeding in the flea population. This ensures that the new generation of fleas has a litter of young rabbits to infest.

SPECIAL DIGESTIONS: PROCESSING PREY

The Boa

To survive, carnivores must find and subdue prey animals, filter feeders must trawl vast numbers of tiny creatures and parasites must become attached to suitable hosts. But these are only first steps, for unless the nutrients that prey animals contain can be efficiently digested, they are useless. In carnivores the food processing business is carried out by different regions of the gut and starts immediately food enters the meat-eater's mouth. The starfish even turns its gut outside its body to start the treatment.

In or around the mouths of most carnivores are teeth or other organs for breaking up food. For most carnivores these are essential because prey animals are unwieldy lumps of meat too large to swallow. So in carnivores, from insects to big cats, the violence of the kill is immediately followed by a period of mastication.

Once inside a carnivore, food has to be stored and digested. The infrequent, large meals implicit in some hunting methods mean that part of the gut must be large and elastic enough to receive these rare feasts. A sea anemone uses its entire body cavity for this task, the vampire bat packs a vast blood meal into a uniquely specialized stomach and a snake such as a boa constrictor, distorted by the mammal it has swallowed whole, is ample evidence of the expansibility of its gut.

When it comes to digestion, animal food is generally easier to deal with than plant food which contains a high proportion of tough substances like cellulose. This means that the small intestine of a carnivore is usually shorter than that of a herbivore of similar size and secretes more fat- and protein-splitting agents or enzymes. Despite the comparative digestibility of animal food, several carnivores have adopted special strategies for eliminating the indigestible parts of their prey. All birds of prey, for example, regularly regurgitate pellets containing bones, insect cuticles, hair and feathers. Some parasites, in contrast, have learned the digestive tricks necessary for breaking down keratin, the tough component of skin, hair and feathers.

The supreme specialists among snakes, the poison-injectors, such as rattlesnakes and vipers, are predators at the pinnacle of snake evolution. Yet equally fascinating are the non-venomous snakes, including boas and pythons, for they still bear the tell-tale marks of the lizards from which all today's snakes are descended.

Boas and pythons kill their prey—usually birds or mammals—by enfolding them in muscular coils of their serpentine bodies and crushing them to death. Only when the prey has expired through suffocation or internal crush injuries is the toothed mouth used to begin the task of swallowing the prey. As in most snakes, the skulls of boas and pythons are modified to provide enormously wide gapes that enable prey much larger than the snake's head to be swallowed. The elastic, capacious gut receives and stores the meal. Rapid digestion starts in the stomach and is finished in the intestine.

Zoologically close relatives, boas and pythons are grouped together in the family Boidae, but live in opposite parts of the globe. The boa constrictor, *Constrictor constrictor*, is found from California to Argentina, while the pythons are inhabitants of the Old World.

The boa constrictor is an adaptable snake in its choice of habitats and its feeding methods. It can make a predatory living in ecological situations ranging from semi-deserts to tropical rain forests and it sometimes hunts in trees—aided by its partly prehensile tail—as well as on the ground. The largest authenticated record for a boa is of a specimen 18 feet 2 inches (5.6 metres) long.

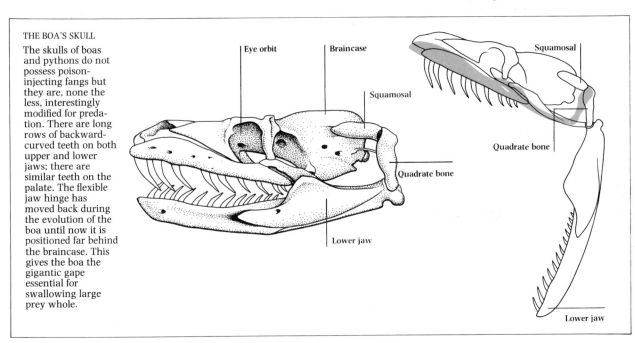

THE BOA'S SKULL
The skulls of boas and pythons do not possess poison-injecting fangs but they are, none the less, interestingly modified for predation. There are long rows of backward-curved teeth on both upper and lower jaws; there are similar teeth on the palate. The flexible jaw hinge has moved back during the evolution of the boa until now it is positioned far behind the braincase. This gives the boa the gigantic gape essential for swallowing large prey whole.

Eye orbit   Braincase   Squamosal

Squamosal

Quadrate bone

Quadrate bone

Lower jaw

Lower jaw

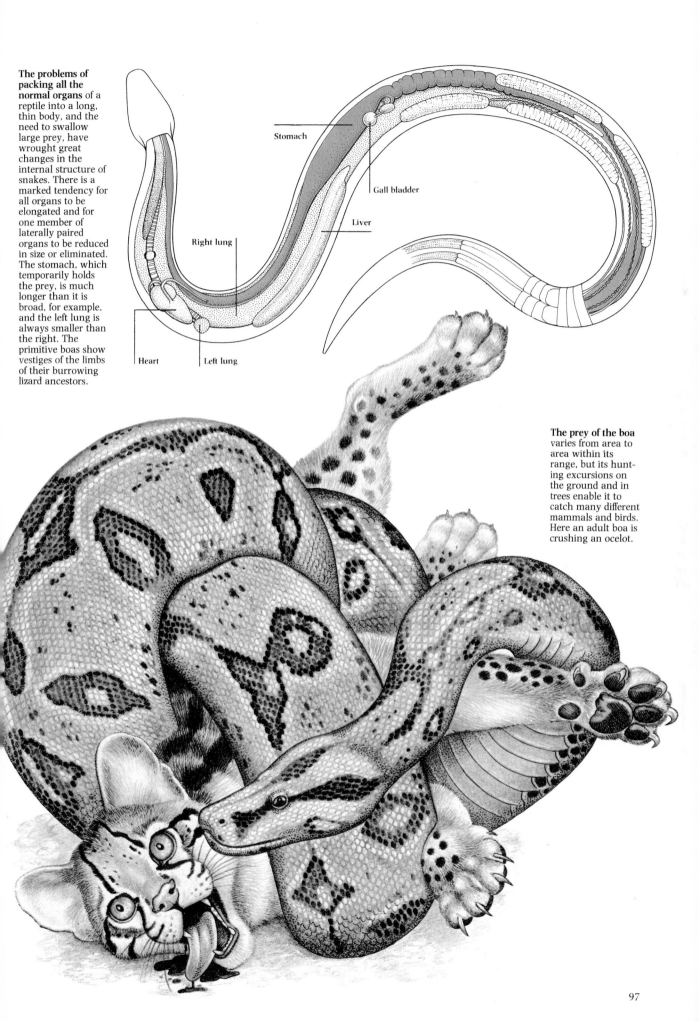

The problems of packing all the normal organs of a reptile into a long, thin body, and the need to swallow large prey, have wrought great changes in the internal structure of snakes. There is a marked tendency for all organs to be elongated and for one member of laterally paired organs to be reduced in size or eliminated. The stomach, which temporarily holds the prey, is much longer than it is broad, for example, and the left lung is always smaller than the right. The primitive boas show vestiges of the limbs of their burrowing lizard ancestors.

Stomach

Gall bladder

Liver

Right lung

Heart

Left lung

The prey of the boa varies from area to area within its range, but its hunting excursions on the ground and in trees enable it to catch many different mammals and birds. Here an adult boa is crushing an ocelot.

The Vampire Bat

By taking to the skies the bats have become mammalian masters of the night. Such nocturnal, airborne superiority has opened up many possible life-styles for these winged, warm-blooded predators and, in the course of their evolution, bats have come to take advantage of almost every food source that can possibly be gathered within their physical and behavioural limitations.

While many bats have retained the primitive, insect-eating habit of their ancestors, others have moved on to catch bigger prey. Specialist bats can take rats from the ground or gecko lizards from trees. Some even hunt osprey-style for fish swimming just below the surface of pond or river water. Yet other groups of bats have become herbivores—the largest of all bats, the "flying foxes", eat fruit.

Of all the bats' various feeding methods, none has stimulated man's imagination more than the technique employed by three bat species of South

and Central America whose very name—vampire bats—conjures up a legend. Except for the fact that they feed on blood, the real vampires are nothing like their fictional counterparts.

The three species of vampire bats are all small, measuring 2 to $3\frac{1}{2}$ inches (5 to 9 centimetres) with a wing span of 5 to 6 inches (12.5 to 15 centimetres). The best studied form is the common vampire, *Desmodus rotundus*. This brown bat, with complex, folded ears and a small, leaf-like nose, is found from northern Mexico south to Chile, Uruguay and Argentina.

During daylight hours, vampires roost together in large numbers in caves, hollow trees or old buildings. They hang upside down in their roosting places using the hooked claws on their hind legs. Soon after darkness falls the bats leave their daytime quarters *en masse* then disperse to locate their prey. Man is only occasionally the vampire's victim; domestic animals such as horses are much more common sources of food.

Like insectivorous bats, vampires find their prey using echo location, but, because a target the size of a horse reflects much more sound than a flying moth, the ultrasonic pulses emitted by a vampire can be effective at much lower power than those of the insect-eaters. Once on the surface of a victim, the vampire uses its teeth to scrape a feeding wound in the skin, then laps and sucks the blood that flows from it. An important feature of the vampire's blood-feeding habit is its secretion of saliva that can inhibit blood clotting, for if the blood in the wound congealed, the feeding period would be severely curtailed.

A vampire bat consumes about 46 pints (26.2 litres) of blood in a year but the blood lost in a single feeding attack by a vampire on a man or a horse would never be enough to do any serious harm to the victim. A much more serious threat is the transmission of rabies. In Mexico it has been proved that vampire bats have passed rabies on to cattle.

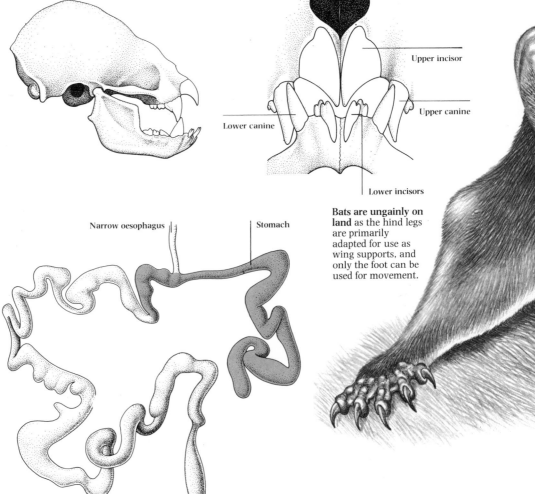

Specialized teeth are the most remarkable feature of the skull. The upper incisors and canines are adapted to shave a shallow wound, down to the skin capillaries of the prey, to feed from. The vampire must do this without disturbing the prey. There is an upper incisor on each half of the jaw; each has a central point and long, sharp edges. The canines are smaller but sharp.

Upper incisor

Upper canine

Lower canine

Lower incisors

Blood is the sole food of the vampire bat. It is an excellent all-purpose diet but with two disadvantages: it clots and it is bulky with a large water content. The gut is equipped to deal with these problems. Salivary glands produce secretions that prevent the clotting of vertebrate blood and the oesophagus is a narrow tube ideal for liquid food. The stomach itself is an elongated bag with a free portion capable of enormous distension during a feeding bout. Gorged bats can be so heavy they can hardly fly.

Narrow oesophagus

Stomach

Bats are ungainly on land as the hind legs are primarily adapted for use as wing supports, and only the foot can be used for movement.

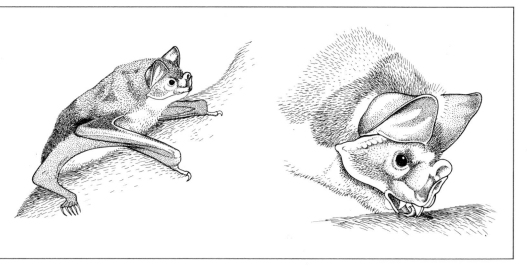

A BLOOD MEAL

The vampire bat finds its prey animal using low power ultrasonic pulses. The victim will usually be asleep and the bat lands either directly on its back or on the ground nearby. It then clambers up the animal to reach a suitable feeding site, often the neck. Resting lightly on its hind feet and forelimb thumbs, the bat shaves a feeding wound and starts its blood meal.

A common vampire bat, *Desmodus rotundus*, has shaved its feeding ground on the prey, allowed the blood to well up and is feeding. Once the blood is flowing, the bat feeds with its muscular tongue, partly lapping, partly sucking the blood. The tongue is grooved and can be extended over a notch at the centre of the lower lip.

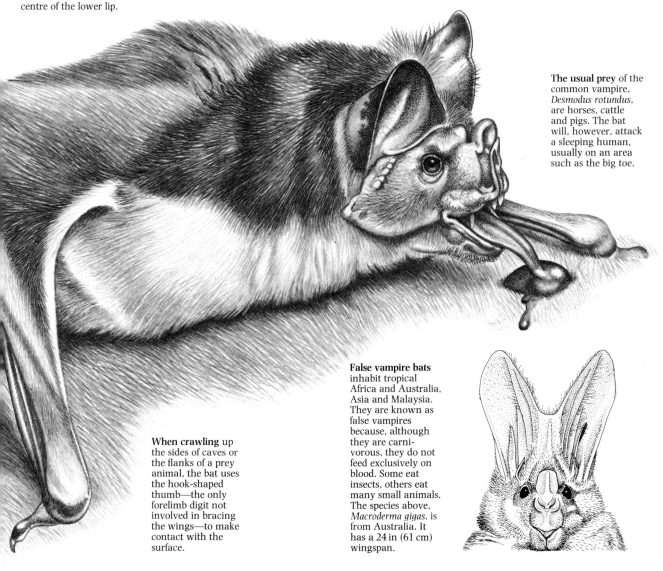

The usual prey of the common vampire, *Desmodus rotundus*, are horses, cattle and pigs. The bat will, however, attack a sleeping human, usually on an area such as the big toe.

When crawling up the sides of caves or the flanks of a prey animal, the bat uses the hook-shaped thumb—the only forelimb digit not involved in bracing the wings—to make contact with the surface.

False vampire bats inhabit tropical Africa and Australia, Asia and Malaysia. They are known as false vampires because, although they are carnivorous, they do not feed exclusively on blood. Some eat insects, others eat many small animals. The species above, *Macroderma gigas*, is from Australia. It has a 24 in (61 cm) wingspan.

The Sea Anemone

With their spectacular white, green, blue or red patterns and circlets of petal-like tentacles, sea anemones have the appearance of harmless underwater flowers. They are, in fact, carnivorous predators and are capable of catching a wide variety of other marine animals.

Each tentacle is a contractile organ studded with nematoblasts, or stinging cells, which trap crustaceans or fish unfortunate enough to swim into, or be carried against them by coastal cross currents. Some stinging cells wrap around protuberances, such as the cuticle spines on a prawn; others drill into prey and inject poisons, still others are so sticky that they force prey to remain attached to the tentacle.

Sea anemones are one of the commonest forms in the group known as the Anthozoa, which also includes many types of coral, sea fans and sea pansies. Sea anemones spend their lives attached to the sea bottom, generally to some hard surface such as a rock. Their tentacles extend their range of feeding activity to compensate for this lack of movement.

These underwater predators are distinguished by wide, cylindrical bodies or columns, one end of which contains glands that allow them to grip tightly to the surface beneath them. At the centre of the other end is the mouth, which

opens via a long muscular tube, the pharynx, into the anemone's blind ending gut, or enteron. The mouth entrance is fringed with long tentacles that carry much of the animal's stinging cell armament.

The entire internal space of the sea anemone is one huge alimentary bag into which food can be pushed through the pharynx. In effect, the gut is also the body cavity and, as such, plays an important role in maintaining the animal's shape. Running from the mouth down one or two strips of the inner surface of the pharynx are ciliated tracts, or siphonoglyphs. These enable the pharynx to act like a combined pump and non-return valve because they continuously push sea water into the gut and keep it expanded at a slightly higher pressure than the surrounding water.

The gut is partitioned by mesenteries that stretch inwards from the outer body wall. Some link the outer wall to the pharynx, others wave freely in the gut space and bear on their thickened inner edges flagella whose constant beating keeps the gut contents stirred. In addition, the mesenteries produce digestive enzymes and often carry more stinging cells. They also serve to increase the food-absorbing surface of the gut, or enteron.

Free-swimming crustaceans such as shrimp, prawns and small crabs, as well as fish, are captured by the opportunist sea anemone. Such is the elasticity of its body wall that the anemone can take prey which appears too large. The sea anemone lives on the turbulent strip of land between high and low tide levels, where many creatures normally attached to rocks are swept off and buffeted around by the currents. Sea anemones can take full advantage of this vulnerable food supply.

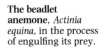

The beadlet anemone, *Actinia equina*, in the process of engulfing its prey.

If attacked by a large bird or fish, the sea anemone reacts by contracting to a conical blob of hard jelly with all the tentacles tucked inside. To contract the body, muscles running down the mesenteries shorten and force water out of the mouth and pores in the body wall of the column.

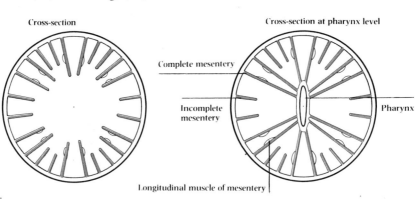

Cross-section

Cross-section at pharynx level

Complete mesentery

Incomplete mesentery

Pharynx

Longitudinal muscle of mesentery

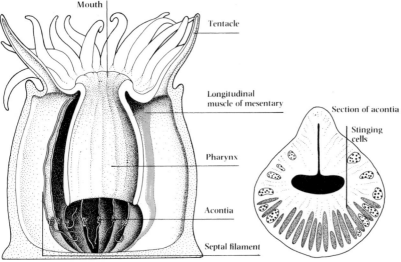

Mouth

Tentacle

Longitudinal muscle of mesentary

Pharynx

Acontia

Septal filament

Section of acontia

Stinging cells

The inner edge of each mesentery is thickened with a rim of tissue, a septal filament, which carries digestive cells. The filament also has rows of stinging cells to deal a final blow to prey. The filaments can extend freely from the mesenteries and hang in the gut as long threads, acontia, covered with stinging cells. An outrush of water during contraction carries the acontia through the body wall, making the anemone unpleasant for a predator.

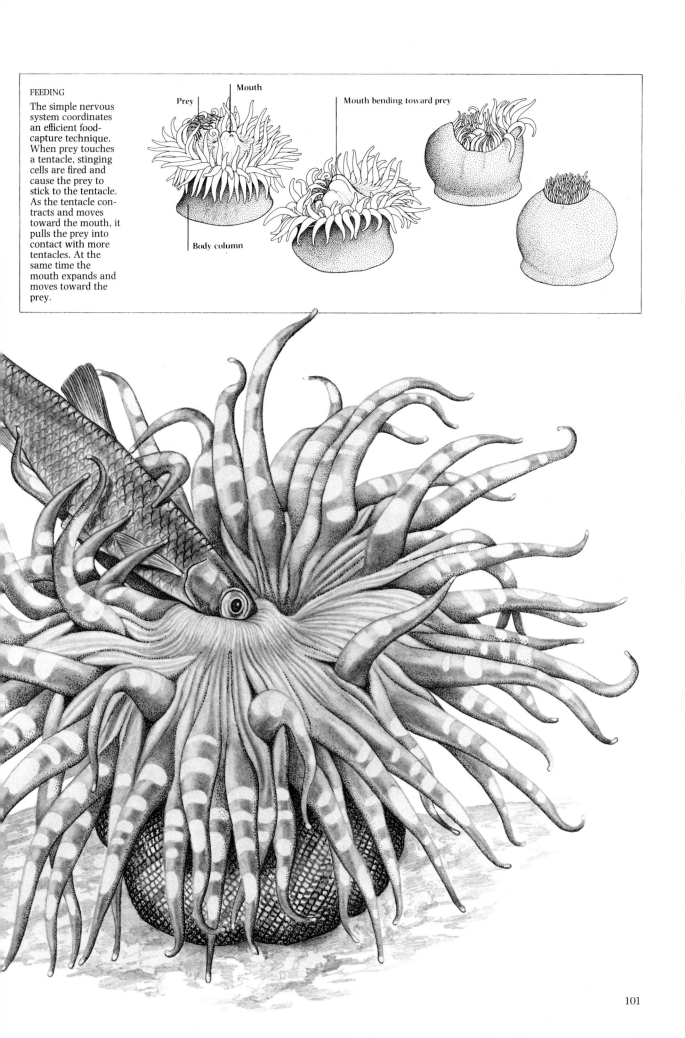

FEEDING

The simple nervous system coordinates an efficient food-capture technique. When prey touches a tentacle, stinging cells are fired and cause the prey to stick to the tentacle. As the tentacle contracts and moves toward the mouth, it pulls the prey into contact with more tentacles. At the same time the mouth expands and moves toward the prey.

Prey

Mouth

Body column

Mouth bending toward prey

EXOTIC WEAPONS: THE ARMS RACE

From the stone axe to the neutron bomb, man has used his ingenuity to produce awesome weapons of destruction. But inventiveness in creating an offensive armoury is not an exclusively human attribute, for the age-old battle between predatory animals and their prey has engendered a precisely similar escalation in the armaments of attack and counter-attack.

If predators are to be successful, they must evolve organs that can subdue and kill prey. Equally, prey animals develop, through evolution, ever more effective defences against the predators' killing techniques, which in turn force the predators to refine their killing weapons. In the survival struggle, competition between predator species that share the same types of prey also pushes individual carnivores along the evolutionary path toward developing the ultimate weapon. In this sector of the battlefield, the evolutionary victors are the predators which kill most prey and therefore live to produce more offspring in future generations.

These two-pronged influences have created a vast array of killing weapons in predators. Complex weaponry is not the exclusive property of the vertebrates, the most recent animals to have evolved. Even lowly invertebrates can have complicated instruments of attack. The corals, sea anemones and jellyfish, for example, have remarkable stinging cells. Packed into each of these cells is a sensitive triggering device, a reservoir of potent toxins and an extensible, drill-like delivery thread which can be thrust out to bore into a prey and inject it with the poisons. These sophisticated killer cells are almost certainly a biological invention nearly 1000 million years old.

Throughout the animal kingdom, some basic methods and organs of attack reappear time and again. Curved, pointed daggers of hard material, for example, which can be used to impale and damage—if not kill—prey, have been developed independently in many animal groups. These daggers come in the form of long claws and as canine or canine-like teeth in a wide range of vertebrates. Birds such as the woodpecker have sharp beaks to use as chisels while carnivorous arthropods employ mouthparts called mandibles to impale prey.

Long limbs tipped with closing claw mechanisms are other common weapon systems. Crustaceans, including crabs and lobsters, use such an arrangement of pincers for offence and defence. Arachnids like spiders, scorpions and pseudoscorpions all have similar armaments. Some specialized insect predators also employ the same technique. An insect destined to make a meal for a praying mantis ends its life ensnared between the closing sides of a pincer-like trap made from the mantid's forelegs set with pointed spines.

Apart from these common weapons, it seems that every biologically possible idea for destruction has been developed among today's predators. From the high voltage, disabling discharges of electric eels to the potent chemical warfare of poisons injected by snakes, wasps and scorpions, the inventiveness of animal armouries seems endless. One marine protozoan concentrates strontium in its body with the skill of a chemical engineer. If a species, as yet undiscovered, can concentrate uranium, let us all beware.

The Electric Eel

The muddy waters of the Amazon and Orinoco river systems of South America are home to the electric eel, *Electrophorus electricus*. These waters are often poor in oxygen and the fish can rise to the surface to gulp air from which it absorbs some oxygen via zones of blood vessels inside its mouth.

This interesting air-breathing ability alone would make an intriguing fish, but pales into insignificance when compared with its other unusual attribute, for the electric eel kills its prey with 550-volt electric shocks.

The electric eel's thick, cylindrical body is largely made up of electric organs—modified muscles which no longer have the power to contract, but which can release high voltage discharges into the surrounding water. Each electric organ is composed of thousands of electroplates stacked face to face. Single electroplates can produce only 0.1 to 0.15 volts, but packed together in large numbers their voltages can be added together to produce biologically damaging discharges.

Today the electric eel uses its electrical abilities to stun and kill fish and amphibians for food. Originally, however, this fish and others like it probably produced discharges of lower power as threats or for object location, just as fish such as *Gymnarchus* still do with precision. The electric eel's equipment bears out this assumption, for it has three sets of electroplates, one large and two small. The large set, whose positive pole lies towards the fish's head, is the killer organ while the smaller pair emit slow pulses of low-power discharge for navigation in murky waters and the detection of underwater objects. When the eel is immobile, only one to five pulses per second are produced, but if it is actively searching for food, up to 50 pulses a second are emitted.

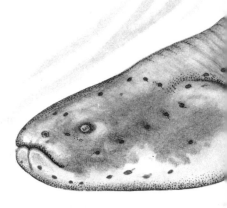

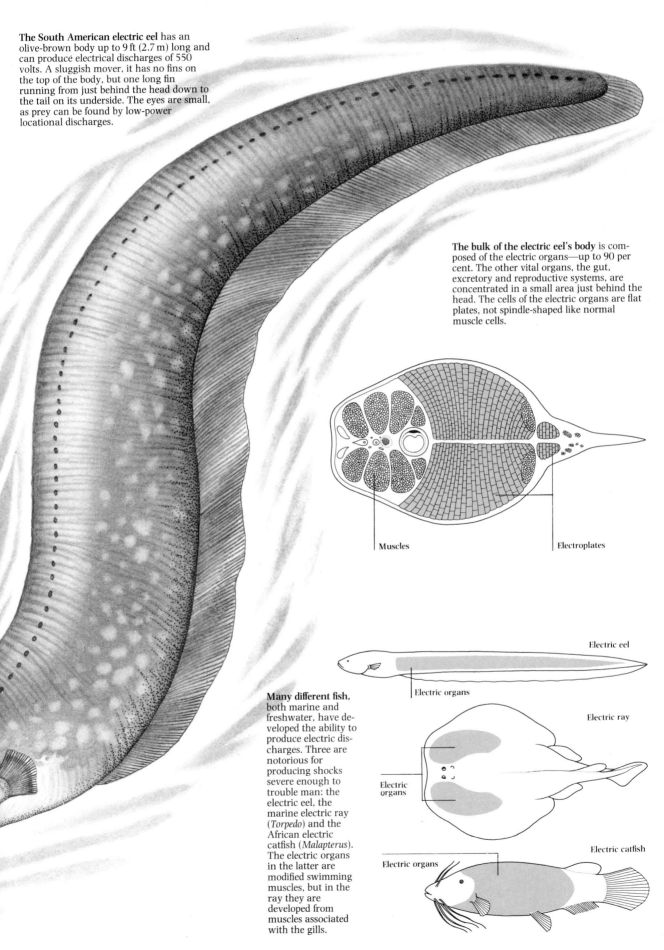

The South American electric eel has an olive-brown body up to 9 ft (2.7 m) long and can produce electrical discharges of 550 volts. A sluggish mover, it has no fins on the top of the body, but one long fin running from just behind the head down to the tail on its underside. The eyes are small, as prey can be found by low-power locational discharges.

The bulk of the electric eel's body is composed of the electric organs—up to 90 per cent. The other vital organs, the gut, excretory and reproductive systems, are concentrated in a small area just behind the head. The cells of the electric organs are flat plates, not spindle-shaped like normal muscle cells.

Muscles

Electroplates

Electric eel

Electric organs

Electric ray

Electric organs

Electric catfish

Electric organs

Many different fish, both marine and freshwater, have developed the ability to produce electric discharges. Three are notorious for producing shocks severe enough to trouble man: the electric eel, the marine electric ray (*Torpedo*) and the African electric catfish (*Malapterus*). The electric organs in the latter are modified swimming muscles, but in the ray they are developed from muscles associated with the gills.

The Octopus

Intelligent, usually fast-swimming animals with superb vision and a devastating array of offensive organs for attacking prey, squid and octopuses are the supreme invertebrate predators. These tentacled creatures surprisingly share a place in classification with the slow-moving slugs and snails: all belong to the phylum Mollusca, in which the squid and octopus are grouped in the class Cephalopoda.

The cephalopods are an ancient class. Forms with external shells are known from the Cambrian era of 500 million years ago and cephalopods have dominated the life of ancient seas in many geological periods since. But although more than 10,000 different fossil cephalopods are known, only some 400 species exist today.

Squid, elongate, streamlined, mid-ocean swimmers equipped with huge eyes to find their prey, are the most direct descendants of the cephalopods of old. Having located fish or other prey animals the squid captures them by rapidly shooting out two long tentacles bearing suckered tips. The food is then hauled to the mouth region where it is held by a ring of eight much shorter tentacles. Speed through the water is achieved by the jet propulsion mechanism. Water is forced like a rocket exhaust from a chamber (the mantle cavity) through an adjustable nozzle (the funnel) which can be aimed to allow the squid to streak in any direction.

The octopus, a hunter of shallow seas, is adapted for a totally different life-style. Ambush is its basic hunting method. Camouflaged by its sophisticated system of colour-changing skin cells, an octopus lurks unseen in rock gulleys or coral reef crevices. When a suitable prey—a fish or more commonly a crustacean—moves within range the octopus shoots out one of its eight long tentacles, armed with many muscular sucking discs, to trap it. Any of the tentacles can be used to ensnare prey and all are sensitive organs.

Unlike squid, the bulbous octopus is not built for speed. A concerted gliding, pulling action of the suckered arms in contact with the sea bed usually moves an octopus along, or it can move in a similar fashion to the squid by jet propulsion. In emergencies it can use its jet propulsion system to squirt out a "smoke screen" of ink secretion into the water and confuse an attacker.

Despite popular myth, octopuses are never very large. The body length of the largest species, *Octopus punctatus*, found on the Pacific coast of the United States, never exceeds 12 inches (30 centimetres), although when extended its tentacles may reach 16 feet (4.8 metres).

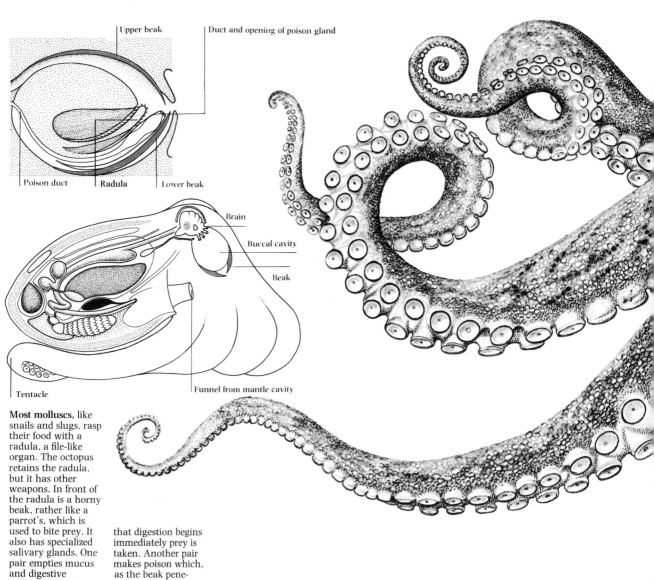

Upper beak

Duct and opening of poison gland

Poison duct　Radula　Lower beak

Brain

Buccal cavity

Beak

Tentacle

Funnel from mantle cavity

Most molluscs, like snails and slugs, rasp their food with a radula, a file-like organ. The octopus retains the radula, but it has other weapons. In front of the radula is a horny beak, rather like a parrot's, which is used to bite prey. It also has specialized salivary glands. One pair empties mucus and digestive enzymes into the radular region, so that digestion begins immediately prey is taken. Another pair makes poison which, as the beak penetrates, is forced into the prey.

The eyes of the octopus and man are built on extremely similar lines. This similarity is due to functional parallels and the constraints imposed by the principles of bio-engineering. It seems that efficient eyes can be built in few ways, one of which is in the form of a camera with a sensitive retina, a focusing lens and an iris.

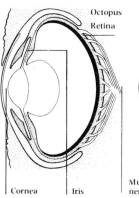
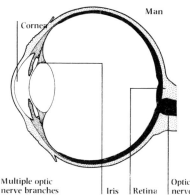

Octopus

Retina

Cornea

Man

Cornea

Iris

Multiple optic nerve branches

Iris

Retina

Optic nerve

The common octopus, *Octopus vulgaris*, hunts in inshore waters. Once the octopus has spotted its prey using its superb eyes, the tentacles are its crucial predatory weapons for grasping the prey and pulling it back to the beak. The double rows of suckers on each tentacle are richly endowed with sense cells. These enable each tentacle to be both a powerful, gripping weapon and a sensitive probing and manipulative organ. Both the male and female octopus have colour-changing abilities. Pigment sacs, operated by radiating muscular spokes which expand or contract the pigment, allow the octopus to make rapid changes in its colour and pattern both for camouflage and as a warning signal.

Crabs and other slow-moving, bottom-dwelling crustaceans are the main items of the octopus's diet. Its powerful beak can easily pierce or crush crab and lobster skeletons.

The octopus has a rounded body and no fins. It is more versatile in the use of its tentacles than other cephalopods, employing them as agile, flexible legs to pull itself along the sea bottom. To make faster, darting movements, the octopus uses the jet propulsion technique. It swims horizontally with tentacles trailing.

Octopus crawling

Octopus swimming

The Crown of thorns Starfish

Found in every ocean of the world and frequently seen along rocky or sandy shore lines, echinoderms are large and distinct enough to have been given a number of common names—sea urchins, starfish, sea lilies, brittle stars and sea cucumbers. These spiny-skinned sea creatures of widely differing appearance, many of them with predatory life-styles, all belong to the same animal group.

Certain characteristics are common to the adult echinoderms. Many of their organ systems are arranged in sets of five. This symmetry usually imposes a radial pattern on much of the body. In addition, they all have a unique set of organs for walking, adhesion and prey manipulation—the tube feet. These are hydraulically operated cylinders of muscles with a mucus-secreting suction cup at their outer ends. There are hundreds of tube feet on the underside of the arms of the crown of thorns starfish, *Acanthaster planci*.

Although present-day echinoderms are totally different from other animals, it is believed that the early forms, or a group of animals very similar to them in organization and development, provided the vital jumping-off point for the line of evolution that leads directly to fish, amphibians, reptiles, birds, mammals and man. For that reason alone, the biology of contemporary echinoderms deserves more than passing attention.

Perhaps the best-known echinoderms are the starfish, or asteroids. Generally five-rayed, these star-shaped animals of the sea bottom and seashore pools are active predators. For more than 2000 years they have been a scourge of oyster beds used by man. Fishermen have been perplexed by their ability to feed on the meat inside oyster shells, particularly since it is so difficult for an inexperienced man to prise apart these shells. The crown of thorns starfish, which feeds on coral polyps, has destroyed many miles of the Australian Great Barrier Reef.

Although the majority of starfish are five-armed, a number of forms have departed from this plan. The sunstar (*Solaster*) may have from 15 to 50 arms, and the crown of thorns has about 20.

The basic body plan of the starfish consists of a central disk with a mouth in the middle of its undersurface. Arms project laterally from this disk. Stomach regions of the gut are in the middle of the disk, but long, blind-ending pouches of gut extend into the arms. The anus is located on the upper side. The soft inner organs of the starfish are protected by a skeleton of crystalline plates just under the body epidermis.

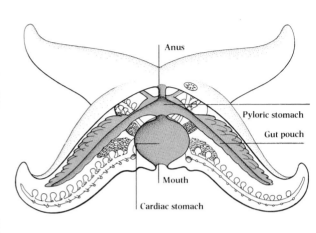

FEEDING
The anatomy of the starfish is essentially radially symmetrical. The gut runs from the bottom to the top of the central disk. The mouth, on the lower side, opens into a cardiac stomach. Above this is the smaller pyloric stomach leading to the rectum and anus. Gut pouches, often 2 to each arm, radiate from the pyloric stomach. They provide a large surface area for food absorption.

Anus

Pyloric stomach

Gut pouch

Mouth

Cardiac stomach

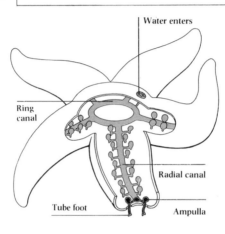

Water enters

Ring canal

Radial canal

Tube foot

Ampulla

Tube feet are operated by a water-vascular system. Water enters through a plate on the upper surface of the starfish. It passes to tubes in the arms and then to reservoirs, ampullae, connected to the hollow tube feet. As the ampulla contracts, it forces fluid into the foot and extends it. The foot muscles squeeze water out again and retract the foot.

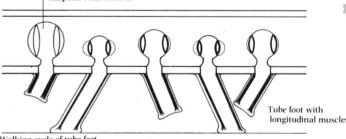

Ampulla with muscles

Tube foot with longitudinal muscles

Walking cycle of tube feet

A killer of oysters. *Asterias* wraps its arms round the bivalve and uses its tube feet to exert a strong pull, perhaps equivalent to $6\frac{1}{2}$ lb (3 kg), on the shells. Within 5 or 10 minutes it opens the shells a fraction of an inch (about 0.1 mm). This is enough to slip its stomach into the oyster and start digestion. In experiments, starfish have even digested the contents of mussels bound with strong wire.

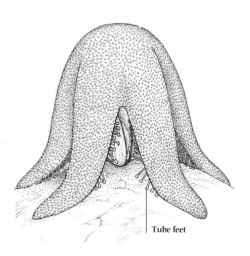

Tube feet

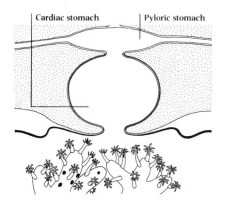

Cardiac stomach	Pyloric stomach

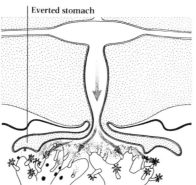

Everted stomach

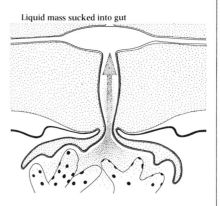

Liquid mass sucked into gut

Most starfish digest their prey externally. The crown of thorns settles itself on a group of its coral prey and, while holding on with the hundreds of tube feet on the underside of its arms, it everts its cardiac stomach over the polyps. Enzymes, produced in the pouches of the gut extending from the pyloric stomach, soon reduce the coral polyps to a liquid mass. The pyloric stomach then seems to pump the food up into the gut. The starfish draws in its stomach and moves on.

The crown of thorns is causing immense damage to coral reefs and at least 100 square miles of the Great Barrier Reef have already been destroyed. Each starfish can measure as much as 16 in (40 cm) across and may consume as much as 16 sq in (103 cm^2) of coral in a day.

The Wasp

Four and a half thousand years ago, so history recalls, Menes, the Pharaoh of Egypt, was stung by a wasp (yellow-jacket) and died. Conflicts between man and wasps are rarely so extreme, but stings from these pugnacious insects are a hot-weather hazard for sunbathers and picnickers alike. In the 50s, 101 people died of wasp stings in the United States.

Wasps get their name from the Anglo-Saxon word "waesp". Each wasp (yellow-jacket) colony has thousands of insects and has a complex social structure based on a fertile queen, a few potent male drones and a vast number of sterile female workers. These workers share in the construction of a multi-chambered nest of plant fibre papier mâché underground, in trees or in buildings. Less well known are the solitary wasps which, although they look like their communal cousins, have

quite different life-styles. Individual solitary females make small nests, usually of mud or clay, containing a few cells for eggs and larvae.

Both social and solitary wasps share the same basic body architecture. The head, carried on a thin neck, is equipped with two large, compound eyes and prominent antennae. The central part of the wasp body or thorax is often hairy and is the attachment area for the six legs, each tipped with double, grasping hooks, and the four wings—a small pair nearer the tail and a larger pair in front.

Behind the wasp thorax, connected via a narrow waist region, is the abdomen which, in many wasps, is striped in yellow and black, although in some species red, orange or metallic reddish hues alternate with the black bands. For social wasps, which use their abdominal stings as reusable defence weapons, this colouring is undoubtedly

intended as a warning. Wasp-attacking predators, if stung, will learn to associate the prominent banding with the discomfort of the sting and perhaps be less likely to attempt an attack the next time. Solitary wasps employ their stings only for paralyzing prey animals so their coloration is most probably a form of mimicry—by looking like their more unpleasant social relatives they gain protection from predators.

All wasps are mixed feeders. The adults eat a great variety of plant-derived foods, of which nectars are especially important, but both solitary and social wasps can be adaptable generalist predators and scavengers or specialist hunters. Solitary wasps stock their nests with paralyzed insects or spiders as food for their larvae. Social wasp workers catch and kill an incredible variety of insects to provide protein for hungry larvae at the nest.

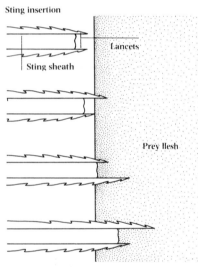

The sting is made of a central sting sheath and 2 pointed lancets. The barbed lancets have grooves along their sides which slide along the edges of the sheath. The 2 lancets are alternately pushed back along the sheath so that the tips "walk" into the prey's flesh.

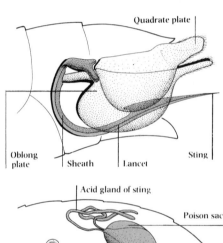

Reproductive and poison organs of a queen wasp

The thrusting action of the sting lancets is achieved by the rocking motion of separate internal plates (the quadrate plate and the oblong plate) to which the sting components are fixed. The wasp hangs tightly on to its victim with its legs and pushes in the sting. The movements of the lancets force venom from the poison sac down the sting shaft and through its tip into the victim. The venom contains protein constituents causing allergic reactions, and histamine, serotonin and kinin which have complex pharmacological effects and probably cause pain.

THE ATTACK

Usually a quick, violent bite from the mandibles will kill the prey; the ultimate weapon of the sting is used only as a last resort. Large prey are often "dissected" so that the most nutritious, and easily carried, part is taken to the nest. Here a wasp kills a butterfly and removes its wings before carrying it off.

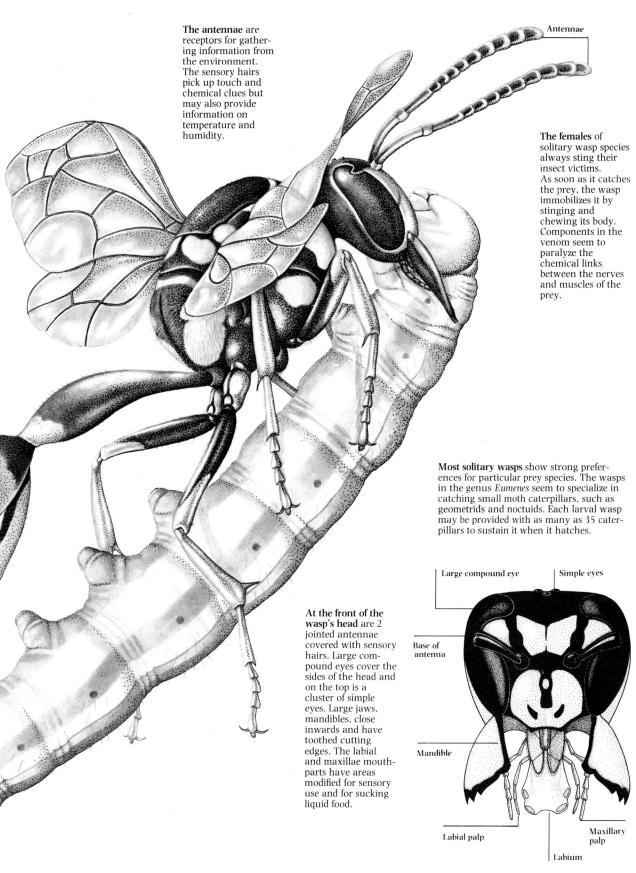

The antennae are receptors for gathering information from the environment. The sensory hairs pick up touch and chemical clues but may also provide information on temperature and humidity.

Antennae

The females of solitary wasp species always sting their insect victims. As soon as it catches the prey, the wasp immobilizes it by stinging and chewing its body. Components in the venom seem to paralyze the chemical links between the nerves and muscles of the prey.

Most solitary wasps show strong preferences for particular prey species. The wasps in the genus *Eumenes* seem to specialize in catching small moth caterpillars, such as geometrids and noctuids. Each larval wasp may be provided with as many as 35 caterpillars to sustain it when it hatches.

At the front of the wasp's head are 2 jointed antennae covered with sensory hairs. Large compound eyes cover the sides of the head and on the top is a cluster of simple eyes. Large jaws, mandibles, close inwards and have toothed cutting edges. The labial and maxillae mouthparts have areas modified for sensory use and for sucking liquid food.

Large compound eye   Simple eyes

Base of antenna

Mandible

Labial palp

Maxillary palp

Labium

Solitary, eumenid wasps like *Eumenes*, above, are active and fearless hunters of insect larvae, such as moth and butterfly caterpillars, but rarely eat them themselves. Adult eumenids feed mainly on flower nectar, and use the prey to feed their young. The female eumenid makes a small nest in which she lays her eggs. The cells or chambers of this nest are filled with prey animals, paralyzed by the wasp's sting, to provide living food for the wasp larvae when they hatch. The species above, the potter wasp, carries hundreds of soft clay pellets from damp soil to the nest site—often a tree—to build a nest like a rounded flask.

The Woodpecker

Insects, spiders and terrestrial crustaceans that inhabit the crevices beneath thick tree bark would appear to be safe from attack by insectivorous birds. Equally, those insects which bore into solid wood and form tunnels in timbers would seem to be fully protected from avian predation. Such conclusions, though, ignore the pressures of selection involved in the battle between predators and prey. One group of birds—the woodpeckers—has become efficient at excavating this apparently invulnerable insect food.

True woodpeckers constitute the bulk of the family Picidae, which contains more than 200 species spread over most parts of the world, including wrynecks, piculets, flickers and sapsuckers. Few species are found in extremely cold climates, as the insect life is too sparse to support them, and there are none in Madagascar, Australia or most of the Pacific islands east of the Philippines and Celebes in Indonesia.

In order to locate prey, almost all woodpeckers climb trees and chisel into bark and wood with their heavy, pointed beaks. Once insects have been exposed by this type of excavation, they are deftly removed by the birds' long protrusible tongues. Some species, however, have learned to gather food in different ways. The sapsuckers, for example, obtain sugar-rich sap from trees that they have "tapped" by chiselling. Their tongues are bristly at the tip, thus enabling them to pick up large quantities of sticky sap. A few true woodpeckers, such as the green woodpecker of Europe (*Picus viridis*), feed primarily on the ground. Probing with their bills, they are adept at catching ants with their immensely long tongues.

Woodpeckers also employ the excavation technique for nest-hole construction. Although many species make their holes in tree trunks, there are others that make tunnels in banks of earth or termite mounds, and still others that bore into large cactus growths. Their rapid, rhythmic blows on the dead branches of a tree, similar to those used in feeding, produce a "drumming" sound. This serves as the equivalent of a territorial call in the breeding season. Since the frequency and length of these staccato drum rolls are often characteristic of a particular woodpecker species, field identification is possible even when the birds have not actually been seen.

The wryneck, of which there are only two species, is believed to represent the ancestral type of woodpecker. So named because of an owl-like ability to twist its neck to a remarkable extent, it perches, instead of clambering up and down tree trunks as do other woodpeckers.

Earwigs, wood lice and some spiders and harvestmen live in crevices under bark flakes and fall prey to the attacking beak and tongue of the woodpecker. Flies, beetles, wasps and their larvae tunnel into wood so are more difficult to find, but provide larger meals.

The strong feet of the woodpecker are specialized for running up and down tree trunks and clinging on while chiselling for food. The 4 toes, each with a large curved claw, are arranged in a zygodactyl pattern: 2 toes pointing forward and 2 backward. The toes of most birds are arranged with 3 forward and 1 back.

Woodpecker

Magpie

The European great spotted woodpecker, *Dendrocopus major*, has black, white and red plumage and is about 9 in (23 cm) long. It is a typical species of old coniferous woodlands. As well as feeding by the normal woodpecker method, it removes seeds from fir cones. It also rings trees, particularly limes, with holes through which it feeds on the nutritious sap.

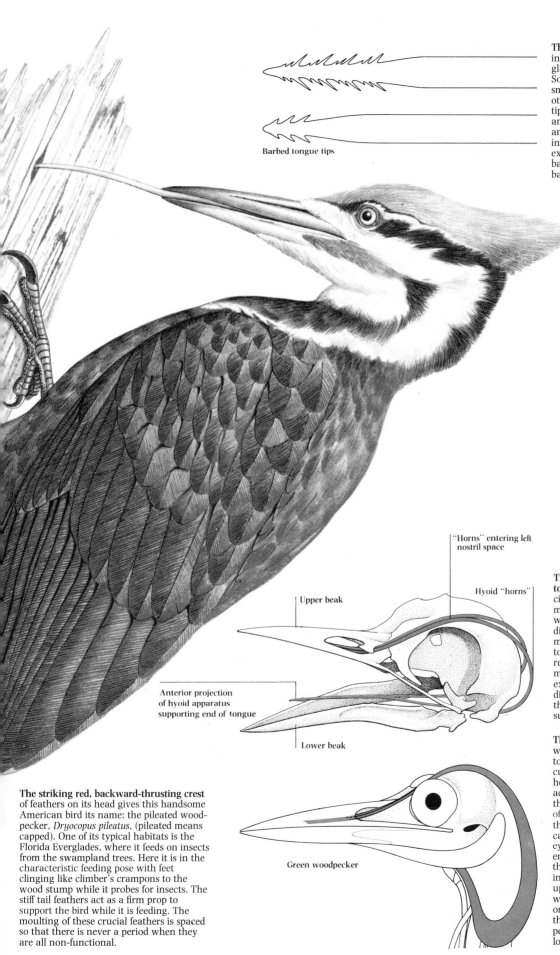

The tongue is coated in sticky mucus from glands at the base. Some species have smooth tongue tips; others have "hairy" tips for sucking sap and picking up ants and other small insects. The most extreme have backward-pointing barbs on the tongue.

Barbed tongue tips

"Horns" entering left nostril space

Hyoid "horns"

Upper beak

Anterior projection of hyoid apparatus supporting end of tongue

Lower beak

Green woodpecker

The specialized tongue and associated bones and muscles make the woodpecker's distinctive feeding method possible. The tongue itself is remarkably long and mobile and can be extruded a great distance because of the length of its supporting bones.

The hyoid bones which support the tongue must be curled up round the head in order to be accommodated in the skull. The bones often curve round the back of the braincase and over the eye orbits. The free ends are located at the base of the bill, in the cavity of the upper beak, or are wrapped round an orbit like a spring. In the green woodpecker the bones loop down the neck.

The striking red, backward-thrusting crest of feathers on its head gives this handsome American bird its name: the pileated woodpecker, *Dryocopus pileatus*, (pileated means capped). One of its typical habitats is the Florida Everglades, where it feeds on insects from the swampland trees. Here it is in the characteristic feeding pose with feet clinging like climber's crampons to the wood stump while it probes for insects. The stiff tail feathers act as a firm prop to support the bird while it is feeding. The moulting of these crucial feathers is spaced so that there is never a period when they are all non-functional.

111

The Scorpion

When animals first began to leave the seas some 300 million years ago to take their chance at life on land, the sting-tailed scorpions were most probably among the pioneers. Fossil records confirm that they were the earliest known land arthropods—joint-legged creatures such as insects, arachnids and crustaceans—to make the transition.

With their relatives the pseudo-scorpions, spiders, harvestmen, mites and ticks, the scorpions are grouped in the class Arachnida. Scorpions vary greatly in size and detailed anatomy from species to species but all today's representatives have a standardized general appearance.

Like the spiders, scorpions have bodies divided into two sections—a forebody, or prosoma, and a hind body, or opisthosoma. Both scorpions and spiders are equipped with two pairs of pincer-tipped grasping organs near their mouths, the chelicerae and the pedipalps, but the pedipalps of scorpions are huge compared with those of spiders.

Although their large pedipalps are used to capture prey, the scorpions have "overkill" in the form of a bulbous poison apparatus and venomous sting carried by every species at the end of the long "tail". The venom produced and injected by this system can kill many sorts of invertebrate prey. Despite the scorpion's notorious reputation, this venom is not usually harmful to man, but a few species do inject poison toxic enough to be fatal to humans. Of these dangerous scorpions the most ferocious are those of the genus *Androctonus*, found in North Africa, and the genus *Centruroides* of New Mexico and Arizona. Venom from one of the North African species can kill a dog in seven minutes and a man in seven hours.

Restricted in distribution to the tropics and subtropics, the scorpions are reasonably abundant today, but only in the Gulf States and southwestern United States do they make up a significant proportion of the invertebrate fauna. In South America they are found as far south as Patagonia. Although scorpions are usually thought of as denizens of desert regions, many species are predators of damp environments such as tropical rain forests. But wherever they do occur, scorpions will usually tolerate only a narrow range of temperatures and humidities. To meet these restrictions they seek out micro-habitats that fulfil their particular needs. In the searing heat of the desert day they hide under stones, dig shallow scrapes or even excavate extensive deep tunnels in pursuit of lower-temperature, higher-humidity soil zones. Only at nightfall do the scorpions emerge to hunt.

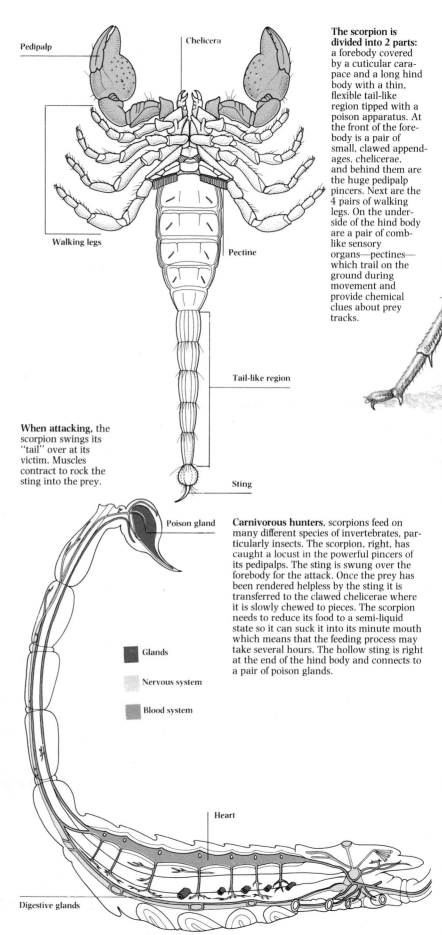

When attacking, the scorpion swings its "tail" over at its victim. Muscles contract to rock the sting into the prey.

Pedipalp

Chelicera

Walking legs

Pectine

Tail-like region

Sting

Poison gland

Glands

Nervous system

Blood system

Heart

Digestive glands

The scorpion is divided into 2 parts: a forebody covered by a cuticular carapace and a long hind body with a thin, flexible tail-like region tipped with a poison apparatus. At the front of the forebody is a pair of small, clawed appendages, chelicerae, and behind them are the huge pedipalp pincers. Next are the 4 pairs of walking legs. On the underside of the hind body are a pair of comb-like sensory organs—pectines—which trail on the ground during movement and provide chemical clues about prey tracks.

Carnivorous hunters, scorpions feed on many different species of invertebrates, particularly insects. The scorpion, right, has caught a locust in the powerful pincers of its pedipalps. The sting is swung over the forebody for the attack. Once the prey has been rendered helpless by the sting it is transferred to the clawed chelicerae where it is slowly chewed to pieces. The scorpion needs to reduce its food to a semi-liquid state so it can suck it into its minute mouth which means that the feeding process may take several hours. The hollow sting is right at the end of the hind body and connects to a pair of poison glands.

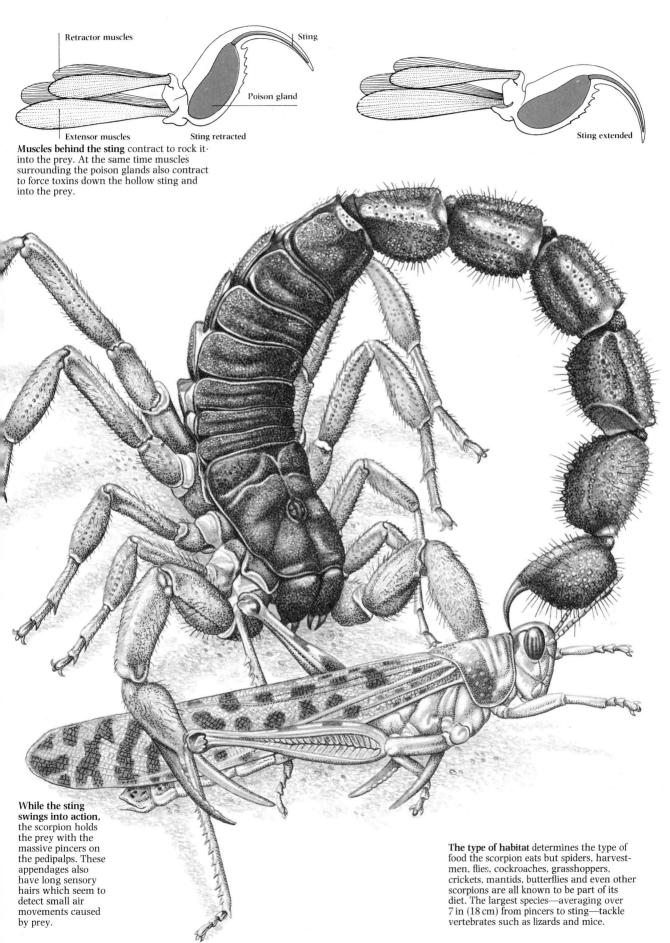

Retractor muscles

Sting

Poison gland

Extensor muscles

Sting retracted

Sting extended

Muscles behind the sting contract to rock it into the prey. At the same time muscles surrounding the poison glands also contract to force toxins down the hollow sting and into the prey.

While the sting swings into action, the scorpion holds the prey with the massive pincers on the pedipalps. These appendages also have long sensory hairs which seem to detect small air movements caused by prey.

The type of habitat determines the type of food the scorpion eats but spiders, harvestmen, flies, cockroaches, grasshoppers, crickets, mantids, butterflies and even other scorpions are all known to be part of its diet. The largest species—averaging over 7 in (18 cm) from pincers to sting—tackle vertebrates such as lizards and mice.

113

TOOLS FOR HUNTING: SIDE-STEPPING EVOLUTION

Man is, of course, the most sophisticated and prolific developer and utilizer of tools—a characteristic that sets him apart from other animals. But in various ways several other higher vertebrates (mammals and birds) have made the transition from an autonomous existence unaided by tools to a life that involves some degree of tool-use. Chimpanzees make use of tools more than any animal other than man.

What is a tool? Strictly speaking, it is an object from an animal's environment that it actively uses to fulfil some function. The concept of active use is a vital part of the definition because if this characteristic is absent from the behaviour then something other than tool-use is being described. Many animals, invertebrate as well as vertebrate, make use of objects in their environment. Marine worms, for instance, build tubes with mud and sand grains; birds construct their nests with grass, twigs and mud. These animals, however, are using construction materials rather than tools. There is also a distinction between tool-use and tool-making. Although a number of animals use external objects for active roles, the object in many cases is in the same form as it exists in the environment.

On the contrary, most of man's tools are artifacts; he modifies the raw materials he finds around him to manufacture something specifically different. A wide variety of animals use tools, but very few actually make them. Even though a mongoose is displaying a remarkably advanced form of behaviour when it throws a stone at a large egg to break it, it would not be true to say that a mongoose had made its tool when it picked up the stone. A chimpanzee, however, will actually strip leaves from a stick to make a probe suitable for extracting termites from their nest. Chimpanzees also chew up leaves to make absorbent sponges for mopping up drinking-water or wiping the body.

An investigation of the ways in which other animals use and construct tools may provide some insights into how man first took the crucial step towards tool-use and then tool-manufacture. The exploratory ways chimpanzees use sticks and stones for a variety of purposes may well be similar to the ways in which prehistoric man used such articles before he began to make tools to a definite pattern.

The most important advantage of tools is that they enable an animal to side-step anatomical evolution. By using a tool, new ways of, for instance, acquiring food can be developed and adapted in a vastly shorter time than it takes to evolve a new set of teeth, claws or tentacles. If the sea otter, for example, had not developed the trick of using a stone as a tool to break the shells of its prey, it would not be able to feed on hard-shelled creatures such as clams and abalones. Thus, tools allow an animal to extend its capabilities for food finding and prey killing and allow it to take advantage of prey species which would otherwise be inaccessible to it. The development of the ability of tool-use depends mainly on two factors: first, a mental capacity with a large enough component of learned behaviour to enable tool-use to become established; second, a marked ability to manipulate objects dextrously and accurately.

The Egyptian Vulture

Birds appear to be poorly equipped to use tools. Their forelimbs are modified solely for flight and their hind limbs are largely specialized for walking or perching. In general, birds manipulate objects in their environment with their beaks, which are capable of exceptionally delicate handling tasks, and their feet. Long-tailed tits or weaver birds, for example, use these two apparently unremarkable appendages to construct their elaborate nests.

Some behaviour patterns displayed by birds, although not involving the use of tools, show a complexity that perhaps presages it. The male satin bowerbird, *Ptilonorhynchus violaceus*, colours the inside of its intricate bowers by applying crushed fruit-pulp or other materials with its beak. The lammergeier, or bearded vulture, *Gypaetus barbatus*, instead of joining other vultures at carcasses, has developed a strikingly unique technique for cracking the large bones of carrion. Since its beak is not powerful enough to crush the bones, the vulture drops them on to a rocky surface while in flight so that it can extract their nutritious marrow once they split. Similarly, a number of gulls drop bivalve molluscs such as mussels on to rocks in order to get at their soft inner parts.

There are two bird species, however, that do make use of tools. The Egyptian vulture, *Neophron percnopterus*, cracks large eggs with stones, and one of Darwin's finches, native to the Galapagos islands, utilizes a cactus spine or small twig, held in its beak, as a slim probe for capturing insects.

The tool-using finch
treats the pointed
cactus spine as a
sharp extension to
its bill. With the help
of the spine it brings
insects in wood or
cactus crevices to
the surface. It then
drops the spine
temporarily while it
picks up the insect
prey with its beak.
Cacti are one of the
dominant plant
forms of the dry
country of the
Galapagos islands.

Darwin's finches, a group of perching birds,
which have evolved and diversified on the
Galapagos islands in the Pacific, exemplify
the adaptability of living creatures. The
basic finch type of the islands has split into
a range of species, each specialized for a
different life-style. One Galapagos finch,
Camarhynchus pallidus, has moved away
from the typical pattern of seed eating and
is largely insectivorous. It improves its
insect-capturing abilities by holding a
cactus spine in its beak to remove prey from
holes. The Egyptian vulture, *Neophron
percnopterus*, is a small black and white
vulture found around the Mediterranean
and into Asia. It is one of the last birds to
arrive at a carcass—bigger carrion-eating
birds, like the griffon vulture, come first. It
also scavenges in villages, taking any edible
garbage it can find.

Most vultures have
areas of bare skin
around the head and
neck region, assumed
to be related to the
bird's messy
carrion-eating habit.
The adult Egyptian
vulture is bald only
at the front of the
face; the back of its
head is covered in
long yellowish
feathers.

**Faced with a large
egg** which it cannot
hold and crush in its
beak, the Egyptian
vulture picks up a
stone and drops or
throws it at the egg
to crack the shell
and get the contents.
Other birds, like the
songthrush, *Turdus
philomelos*, use a
stone as an anvil on
which to break the
shells of snails.

The Chimpanzee

It is impossible to watch the behaviour patterns, social interactions and problem-solving abilities of a group of chimpanzees and not be struck by their similarities to man. The similarity lends weight to the assertion that man is a naked, intelligent, socialized ape.

Considering the close evolutionary proximity of the tail-less ape and man, it is surprising that until very recently so little was known about the ways of the chimpanzee in the wild. Although the antics and tricks of captive ones are intriguing, the natural behaviour of troops of chimpanzees in their native habitat is even more interesting and rewarding to investigate.

A great deal of the present detailed knowledge about the social behaviour of a troop, its food types and tool-use in the wild, stems from the work of a woman biologist and ethologist, Jane Goodall. For ten years she continuously watched a single group of chimpanzees (the Gombe Stream troop in Tanzania) until they became completely used to her presence. As a result, she was able to record their group behaviour.

There is just one species of chimpanzee (*Pan troglodytes*), and it inhabits the humid rain forests of central Africa, from the Niger basin south to Angola. Living in small parties of up to 40 individuals, chimpanzees are adaptable opportunist feeders in the lower reaches of the forest. They eat plants, insects and also hunt and kill other mammals. When erect, the adults stand 3 to 5 feet (90 centimetres to 1.5 metres) high. On the ground, they usually run on all fours, but they are also adept at running with three limbs while holding food in one hand, and at walking on two legs. In trees, these great apes are expert climbers. The chimpanzee makes and uses tools to help it obtain food more than any other animal except man. They use sticks as weapons, for throwing and for probing for insects. Leaves and stones are also employed.

Although chimpanzees are social primates, the structure of a party is not as rigid as the strict "pecking order" of, for example, a baboon troop. Young chimpanzees stay with their mother for several years before they become physically mature. Mature males engage in noisy displays, which partly determine the social hierarchy of male dominance. A dominant male, however, does not have absolute mastery over lower males. Food, for instance, is regarded as sacrosanct and a low-ranking male can protect his own food from a more elevated male.

Within the loosely structured society of the chimpanzee group, promiscuous matings are the rule. When a female comes on heat, several males will gather round and mate with her. Parental, especially maternal, care and feeding continues for several years.

Leaves are used by chimpanzees for food as well as a variety of other functions, such as dabbing bleeding wounds or wiping mud or food remains from their bodies. The chimpanzee also chews clumps of leaves to make an absorbent sponge. With this it can mop up rain water from crevices otherwise inaccessible, and squeeze the water into its mouth. It can also wipe out food fragments from the skulls or carcasses of prey.

The insect-hunting chimpanzee first picks up sticks about 2 to 3 ft (60–90 cm) long from the forest floor or breaks them off trees. It then strips off the leaves to produce a supple probe which it sticks into ant or termite mounds. It withdraws the probe and licks off the attached insects.

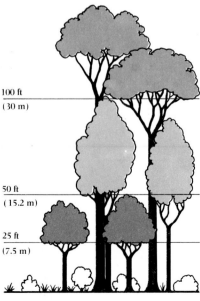

The tropical rain forest is probably the most complex assortment of plant life. A mixture of evergreen trees, creepers and bushes provides a tower of vegetable life and the topmost canopy layers soar up to over 100 ft (30 m). The animals that make use of this multi-storey habitat tend to restrict themselves to vertical zones where they concentrate their hunting activities. Chimpanzees, able to move on the ground and to climb expertly, are usually restricted to the lower zones.

100 ft (30 m)

50 ft (15.2 m)

25 ft (7.5 m)

The intelligent predatory behaviour of chimpanzees and their advanced state compared with non-primate carnivores are illustrated by the way they handle the problem of removing insects from their nests or hiding places. While the aardvark smashes into a termite mound and removes insects with its specialized tongue, the chimpanzee relies on intelligence and manipulative skill and dexterity.

Like man, the chimpanzee is an omnivorous feeder; it eats over 90 different species of trees and plants, while insects, eggs and baby birds provide it with small amounts of animal protein. The most remarkable revelations are those concerning the importance of large prey animals in the chimpanzee's diet. A group of about 40 chimpanzees ate over 20 different types of prey animal in a year, including the young of bushpigs, bushbucks and baboons.

The killing and eating of a young baboon was witnessed by Jane Goodall. As a baboon troop moved through nearby, a chimpanzee grabbed a young baboon. Standing upright and holding the baboon by the legs, it swung it up over its head and smashed the baboon's head on rocks. The killer then sat up in a tree away from the angry baboons and began to feed, ripping the soft flesh from the baboon's belly.

Nests for nighttime "roosting" and hunting bases are made in the lower levels of the rain forest. They are built from branches which are bent and intertwined to form a cup-shaped platform and then lined with twigs.

The Sea Otter

Nineteen species of otter make up the mustelid sub-family Lutrinae. They form a group of aquatic mammals of generally similar appearance. All are long-bodied, short-legged animals with thick, tapering tails. Their heads are flattened from top to bottom with broad, whiskered muzzles. Small ears are almost hidden in the beautiful silky brown fur that completely covers their bodies. A vital part of their swimming equipment, the thick fur consists of a tightly packed under layer that is waterproof and an outer layer of longer, stiffer guard hairs.

Inhabiting fresh water and in some places sea water, otters are active underwater predators. They are superbly efficient, graceful swimmers on the surface, and can be readily identified by a characteristic series of low humps above the water—head, humped back

and end of the tail. Their propulsion is provided by both the undulating movements of their bodies and tails and the powerful kicking of their webbed hind legs. They use their front legs for paddling or else hold them in a streamlined position alongside their bodies. During feeding dives under water, otters swim either with their hind legs or with all four legs trailing passively while they make use of body undulations.

The common, or European, otter, *Lutra lutra*, is found in freshwater rivers and lakes across Europe and parts of Asia to Japan and the Kurile Islands. In times of food scarcity, however, it hunts in coastal marine water. A typical adult measures about 4 feet (1.2 metres) long, including its tail, and weighs 20 to 25 pounds (9 to 11.3 kilogrammes).

Analyses of many thousands of faecal

scats of the common otter in both southern Sweden and Britain have shown that fish constitute at least 60 per cent of its diet. Slow-moving types such as eels are the most commonly preyed upon. Although game fish such as trout and salmon are greatly appreciated by otters in captivity, these fast-swimming fish usually evade capture in the wild.

A few species show marked differences from the common otter. The giant otter, *Pteronura brasiliensis*, which lives in the major river systems of South America east of the Andes and as far south as northern Argentina, is the monster of the otter family. Almost 7 feet (2.1 metres) long, including the tail, it feeds on fish, small mammals and fowl.

Most unusual of all is the sea otter, *Enhydra lutris*, an entirely marine species that is also one of the few animals to use tools.

The skulls of all otters are broad and flattened compared with many other carnivorous vertebrates and the facial portion, the region in front of the eye orbits, is short. The sea otter has a row of 6 incisor teeth in the upper jaw and 4 in the lower. Combined with the pointed canine teeth, these make effective weapons for biting into prey. The cheek teeth, with massive crowns and blunt, flattened cusps, provide a broad area for crushing the hard exteriors of the sea otter's invertebrate prey.

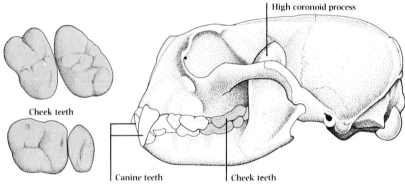

Cheek teeth

High coronoid process

Canine teeth   Cheek teeth

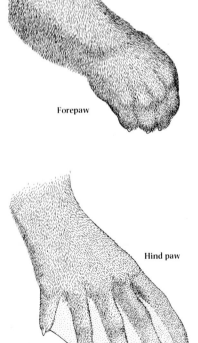

Forepaw

Hind paw

The forepaws are sensitive and manipulative. They each have 5 toes, which are almost fused together, each with a short claw. There is no fur on the palm. The forepaws are used for holding prey and tool stones and for feeling for food in murky water. The hind limbs are larger, more muscular and fully webbed for swimming power.

The sea otter usually dives down to about 33–82 ft (10–25 m) to an area of sea bottom rich in shell-fish or sea urchins. It locates food by sight and touch, picks up both the prey and a stone and returns to the surface. Lying on its back, the otter bangs the prey on the stone to break it open.

Almost exclusively marine, the sea otter rarely comes ashore. The adult male is up to 5 ft (1.5 m) long and weighs up to 80 lb (36.3 kg). The fur varies from black to dark or reddish brown.

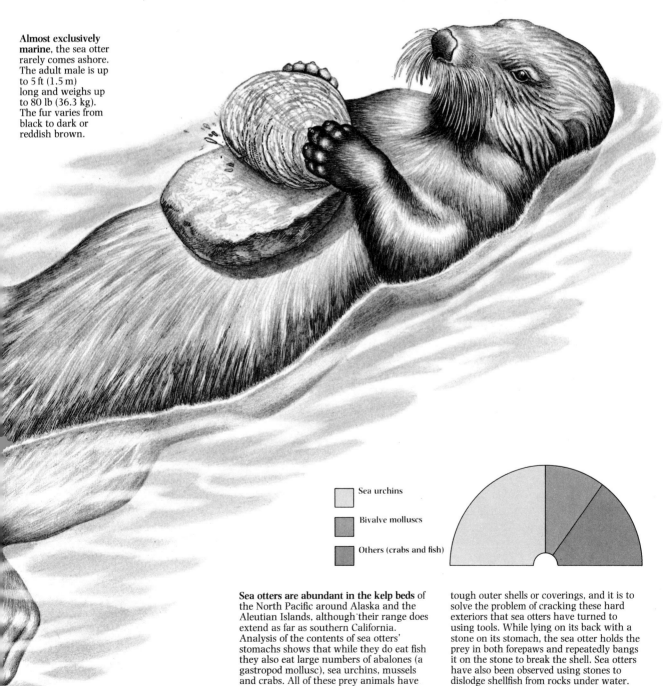

Sea urchins

Bivalve molluscs

Others (crabs and fish)

Sea otters are abundant in the kelp beds of the North Pacific around Alaska and the Aleutian Islands, although their range does extend as far as southern California. Analysis of the contents of sea otters' stomachs shows that while they do eat fish they also eat large numbers of abalones (a gastropod mollusc), sea urchins, mussels and crabs. All of these prey animals have tough outer shells or coverings, and it is to solve the problem of cracking these hard exteriors that sea otters have turned to using tools. While lying on its back with a stone on its stomach, the sea otter holds the prey in both forepaws and repeatedly bangs it on the stone to break the shell. Sea otters have also been observed using stones to dislodge shellfish from rocks under water.

119

FILTER FEEDERS: SMALL IS BOUNTIFUL

Filters—whether they be in a car engine, a laboratory funnel or a fisherman's net—demand a combination of two simple elements: a sheet of material pierced by holes and a fluid or gas that will move through the mesh. This sheet may be solid, jelly-like or woven, the holes minute or large, circular, irregular or slot-shaped. Fluid can move through the filter in two ways—the filter can be pushed through the fluid, or the fluid forced through its holes. Objects in the fluid flow that are smaller than the pores will pass through, while those that are larger than the holes will be trapped on one side of the filter and become concentrated there. But the filter pores will become clogged and the filter less efficient unless the accumulation of particles is continuously or periodically cleared.

These filtering mechanics are vital to the survival of a huge number of predatory animals. Creatures ranging from the smallest, single-celled protozoans to the baleen whales, the largest aquatic vertebrates, catch their prey by filtering. And whatever their absolute size, all the filter feeders trap vast hoards of small prey animals rather than a few large ones. Compared with their own body bulk, all filter feeders process vast volumes of fluid—nearly always water. This is essential to bagging a worthwhile catch because the concentration of food organisms in the water is usually low.

Biological filters vary enormously in origin, shape and size. Some are parts of the animals themselves and branched tentacles, gill walls, bristled antennae and specialized beaks are just a few examples of structures pressed into service as filtering devices. Whatever their origin, however, all these filters consist of repeated units, between which are the holes or interstices that are the filter pores. Almost invariably the holes are elongated slits. Compared with other designs, filters of this sort are easier to construct from living material, are less easily clogged and often provide a high ratio of holes to solid areas. Once caught, food may be moved to the animal's mouth with the help of the filter parts themselves.

Sometimes the filter is specially produced by the animal rather than being an anatomical adaptation. The annelid worm *Chaetopterus*, for example, secretes a mucus bag via which it filters sea water for tiny planktonic micro-organisms. The minute filter pores are readily choked, so periodically *Chaetopterus* bundles up and eats both bag and contents before secreting a new bag. A spider's web, produced from secreted silk, is also a constructed filter: a web, situated in the open and exposed to air currents and winds, filters out any edible insects.

To produce the fluid flow essential to filtering, creatures like sharks, flamingoes and shoveller ducks move their filters through the water. Static or sessile animals depend on the movement of water past their filters, but these organisms can use limbs or other organs to create a water flow. Thus the barnacle waves the water with feather-like limbs, while the sea squirt pumps a stream of water through its body. Such active techniques have the advantage that the animal, by producing its own currents, is released from passive dependence on the whims of the water currents to provide it with food.

Filter Mechanism

The diversity of filter feeding **strategies** adopted by aquatic animals **has** a certain biological pattern. A vast variety of filtering methods are found among the many-celled invertebrates, while tiny single-celled invertebrates—the protozoans—and vertebrates employ a restricted number of mechanisms.

Among the protozoans, only two groups have members committed to a filtering life. These protozoans, the flagellates and ciliates, are named from the cell organs or organelles—flagella and cilia respectively—that enable them to produce efficient water flows for filter-

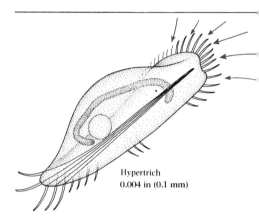

Hypertrich
0.004 in (0.1 mm)

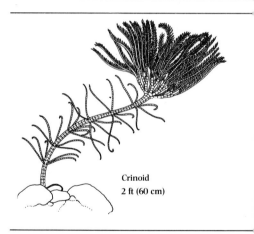

Crinoid
2 ft (60 cm)

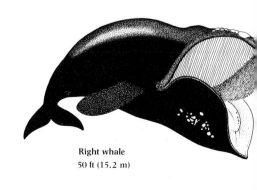

Right whale
50 ft (15.2 m)

ing. Both types of organelle are motile, thread-like extensions of the cell wall supported by a complex scaffolding of narrow tubules, each only 20 millionths of a millimetre in diameter. The activities of these tubules enable the cilia or flagella to bend and to move water, the animal or both. One group of flagellates, the choanoflagellates, have physical filters made up of a "fence" of cell extensions, while in ciliate filter feeders such as *Stentor*, the chonotrichs and hypotrichs, cilia trap some food particles while others are concentrated in eddies of water near the cell mouth.

Only a few vertebrates depend on filtering for survival. Large whales and certain sharks have such a life-style, as do some aquatic birds, notably the ducks and flamingoes. In all cases the filter is in the mouth, gills or beak.

Among many-celled invertebrates, in contrast, there are very few organ systems that have not been adapted for filtering at some time in evolutionary history. The crinoids, for instance, catch prey on their waving arms; rotifers make use of cilia arranged on a food-trapping crown and bivalve molluscs like clams use their gills for filtering food.

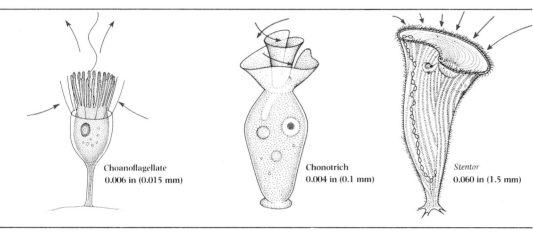

Choanoflagellate
0.006 in (0.015 mm)

Chonotrich
0.004 in (0.1 mm)

Stentor
0.060 in (1.5 mm)

Protozoans
Ciliate filterers like hypertrichs (*Euplotes*), chonotrichs (*Spirochona*) and *Stentor* use either rows of single cilia or cilia fused into undulating sheets to produce a water flow that can be filtered. Other cilia trap the food particles. Choanoflagellates have a flagellum to pull water through a "fence" of cell projections which traps food particles.

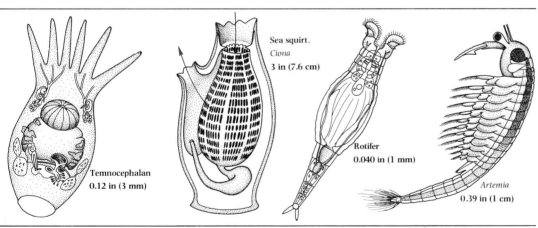

Temnocephalan
0.12 in (3 mm)

Sea squirt,
Ciona
3 in (7.6 cm)

Rotifer
0.040 in (1 mm)

Artemia
0.39 in (1 cm)

Invertebrates
Crinoids have branched arms which trap food particles. Temnocephalans live on animals such as crayfish and filter food with their tentacles from the water flow over the crayfish gills. *Ciona* uses its pharynx region as a filter. Rotifers (*Philodina*) filter water using cilia. The brine shrimp, *Artemia*, filters food with parts of the limbs.

Shoveller duck
20 in (51 cm)

Whale shark
38 ft (11.5 m)

Vertebrates
Baleen whales, such as the right whale, filter feed using huge brush-like plates in the mouth. The shoveller duck has an enormous beak with internal plates (lamellae) to filter its invertebrate food from the water. The whale shark, the largest fish in the world, has gill rakers on its gill arches which strain its food, usually tiny, planktonic crustaceans, from the water.

121

The Humpback Whale

The largest animal that has ever lived on this planet is the blue whale, *Balaenoptera musculus*. It averages 82 feet (25 metres) in length and weighs up to 134 tons. These figures are so greatly in excess of other animals' vital statistics as to make them difficult to visualize. Perhaps it would be more meaningful to point out that one blue whale weighs as much as four of the largest extinct brontosauri, or 30 African elephants, or 1600 average men.

Whales have made drastic modifications to their bodily architecture and physiology in order to become intricately adapted marine predators. Despite their fish-like shape and swimming technique, they have retained the habit of air breathing at the surface of the water.

To sustain their vast bulk of living material, the toothed whales, the Odontoceti, employ orthodox hunting methods. Killer whales, for example, prey mainly on dolphins and seals; sperm whales will dive to 3280 feet (1000 metres) and will remain under water for more than an hour to catch giant squid.

All the non-toothed, whalebone whales, the Mysticeti, have another method of feeding. Plates of whalebone or baleen hang down from their toothless mouth cavities and form a gigantic sieve for straining planktonic invertebrates from the sea water.

Many whalebone whales consume enormous amounts of krill—small shrimp-like crustaceans—such as *Euphausia superba* which occur in dense populations in Arctic and Antarctic waters.

The filter feeding whalebone whales can be divided into two distinct categories: the right whales and the rorquals. The former were so named by early whalers because they were the "right" whales to hunt by traditional techniques. They swim slowly and float when dead. Their vast heads are filled with whalebone plates that can be as long as 10 feet (3 metres). Right whales swim with their mouths constantly open, straining food as it comes into contact with the inner surfaces of the baleen plates.

The rorquals are the giant greyhounds of the whale world. More elongate, streamlined and faster swimming than the right whales, they filter plankton in a discontinuous fashion, opening and closing their mouths.

The humpback whale, a rorqual about 50 feet (15.2 metres) long, inhabits most of the world's oceans. Characterized by long flippers and skin nodules around the head and flippers, it often makes regular long-distance migrations to warmer waters in winter.

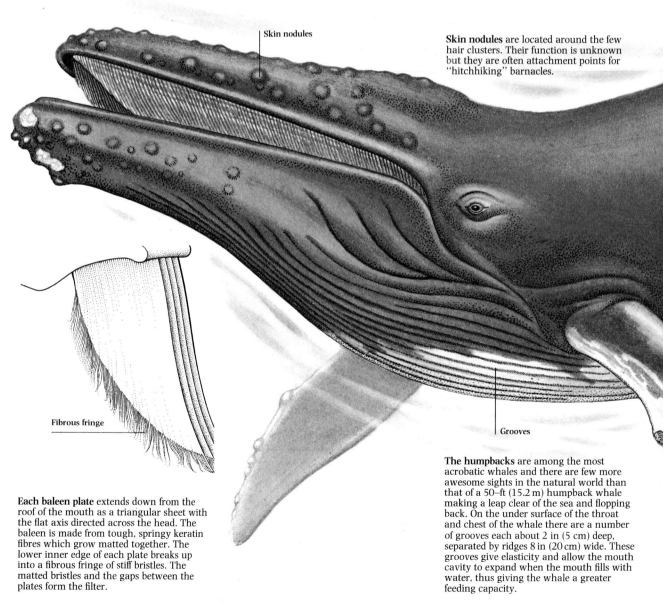

Skin nodules are located around the few hair clusters. Their function is unknown but they are often attachment points for "hitchhiking" barnacles.

Skin nodules

Fibrous fringe

Grooves

Each baleen plate extends down from the roof of the mouth as a triangular sheet with the flat axis directed across the head. The baleen is made from tough, springy keratin fibres which grow matted together. The lower inner edge of each plate breaks up into a fibrous fringe of stiff bristles. The matted bristles and the gaps between the plates form the filter.

The humpbacks are among the most acrobatic whales and there are few more awesome sights in the natural world than that of a 50–ft (15.2 m) humpback whale making a leap clear of the sea and flopping back. On the under surface of the throat and chest of the whale there are a number of grooves each about 2 in (5 cm) deep, separated by ridges 8 in (20 cm) wide. These grooves give elasticity and allow the mouth cavity to expand when the mouth fills with water, thus giving the whale a greater feeding capacity.

Baleen whales, like the humpback, *Megaptera novaeangliae*, lead entirely aquatic lives. The basic mammalian skeleton has been modified to form a massive, fish-shaped structure and a framework for supporting the filter mechanism. Behind the skull, the neck vertebrae are short and often fused together to produce a smooth outline. Tapering flippers have developed from the forelimbs. The pelvic girdle and hind limbs are reduced to minute, non-functional vestiges. The horizontal tail flukes are a novel evolutionary development at the end of the tail vertebrae and have no relationship to the remains of the hind limbs.

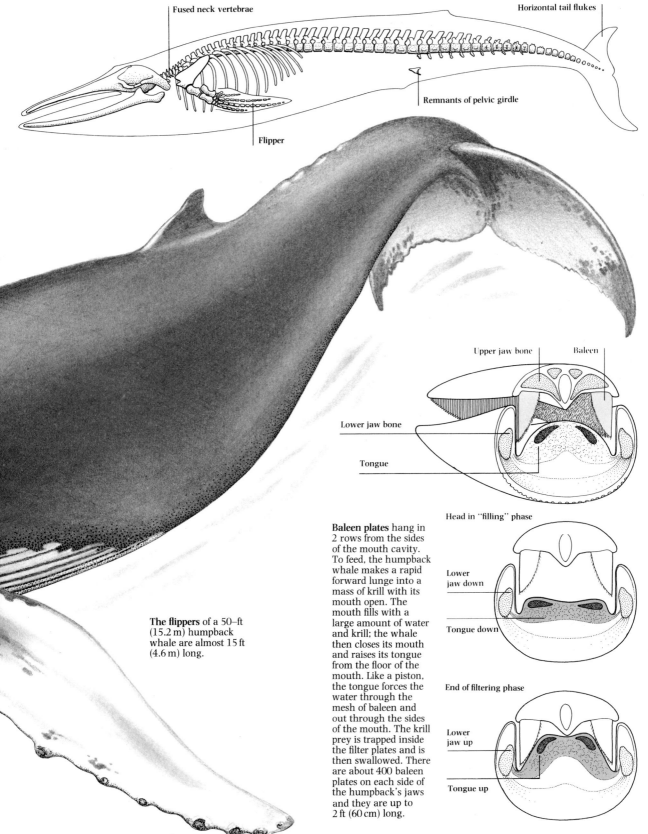

Fused neck vertebrae

Horizontal tail flukes

Remnants of pelvic girdle

Flipper

Upper jaw bone

Baleen

Lower jaw bone

Tongue

Head in "filling" phase

Lower jaw down

Tongue down

End of filtering phase

Lower jaw up

Tongue up

The flippers of a 50-ft (15.2 m) humpback whale are almost 15 ft (4.6 m) long.

Baleen plates hang in 2 rows from the sides of the mouth cavity. To feed, the humpback whale makes a rapid forward lunge into a mass of krill with its mouth open. The mouth fills with a large amount of water and krill; the whale then closes its mouth and raises its tongue from the floor of the mouth. Like a piston, the tongue forces the water through the mesh of baleen and out through the sides of the mouth. The krill prey is trapped inside the filter plates and is then swallowed. There are about 400 baleen plates on each side of the humpback's jaws and they are up to 2 ft (60 cm) long.

The Salp and the Barnacle

Of all the world's many groups of free-living invertebrates, a high proportion have members who make a living as filter feeders. So widespread is the filter feeding habit that some groups of invertebrates are made up almost entirely of animals that filter food from their watery environments. The barnacles are superb examples of static invertebrate filterers, while the salps are equally impressive motile marine feeders.

Although little known outside the province of marine biology, salps are intriguing creatures for, like the sea squirts, they are about as close to being vertebrates as invertebrates can be. The larvae of these animals betray their place in the scheme of zoological kinship, for they possess a skeletal notochord, forerunner of the backbone, but this stiffening rod is never transformed into a bony or cartilaginous spine as it is in true vertebrates.

Deceptively, the adult salp is an in-substantial, jelly-like creature whose whole body is rather like a ram-jet. It is essentially like a barrel-shaped tube through which sea water is forced to produce both a jet propulsion system and the fluid flow vital to filter feeding. Never more than about 12 inches (30 centimetres) long, and often less than 1 inch (2.5 centimetres) from end to end, the salp's most prominent external features are the hoops of muscle that encircle its body partially or completely. These hoops are important to filtering, for they squeeze water in through the water entrance at the head end of the animal—the oral siphon—and out through the posterior "exhaust" or atrial siphon. The water is filtered as it passes through the slot-shaped holes between the delicate plates or lamellae that compose the salp's gills. Trapped food is then transferred to the mouth of the salp.

While salps swim as they filter, adult barnacles never move their bodies as they strain their food. Barnacles are a highly successful group of creatures, for more than 800 species inhabit today's oceans. Fully grown barnacles come in two basic designs. The goose barnacles, such as *Lepas*, attach themselves by long, flexible stalks to any floating objects such as logs, ships or whales; humpback whales almost always have flourishing goose barnacle populations under their chins. Acorn barnacles, such as *Balanus*, settle down on wave-buffeted sea-shore rocks and pilings by means of a broad base; they are shaped like low pyramids to withstand the force of waves and currents.

Both goose and acorn barnacles have exteriors of hard, calcareous plates. Inside this protective armour the soft-bodied crustacean lies on its back and filter feeds. This it does by creating water currents with its jointed, feather-like limbs which are thrust out through a trapdoor at the top of the barnacle to collect food particles.

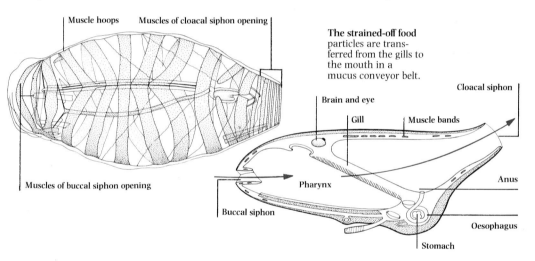

The tube-shaped body of the salp is covered with a transparent, cellulose-containing jacket. The oral opening at the front opens into a pharynx and water passes from this into the posterior (atrial) chamber via 2 gills that act both as a respiratory surface and as a food catching filter. Muscle hoops squeeze water out of the atrial chamber to produce a propulsive water jet.

Muscle hoops   Muscles of cloacal siphon opening

Muscles of buccal siphon opening

The strained-off food particles are transferred from the gills to the mouth in a mucus conveyor belt.

Brain and eye

Gill   Muscle bands

Cloacal siphon

Anus

Pharynx

Buccal siphon

Oesophagus

Stomach

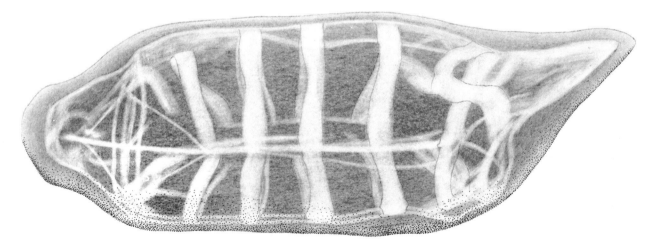

The filter feeding salps are important members of the zooplankton community, often occurring in large numbers. They vary in size but *Salpa maxima*, above, can be as long as 9 in (23 cm). Water currents passing through the body enable the salp to move by a form of jet propulsion, but are also used for filter feeding and respiration. Salps are characterized by their incomplete body hoops and 2 large gill slits. They are solitary animals, but other members of their class, the pyrosomideans, are colonial. These animals are known as "fire bodied" because if touched the whole colony emits a glowing light.

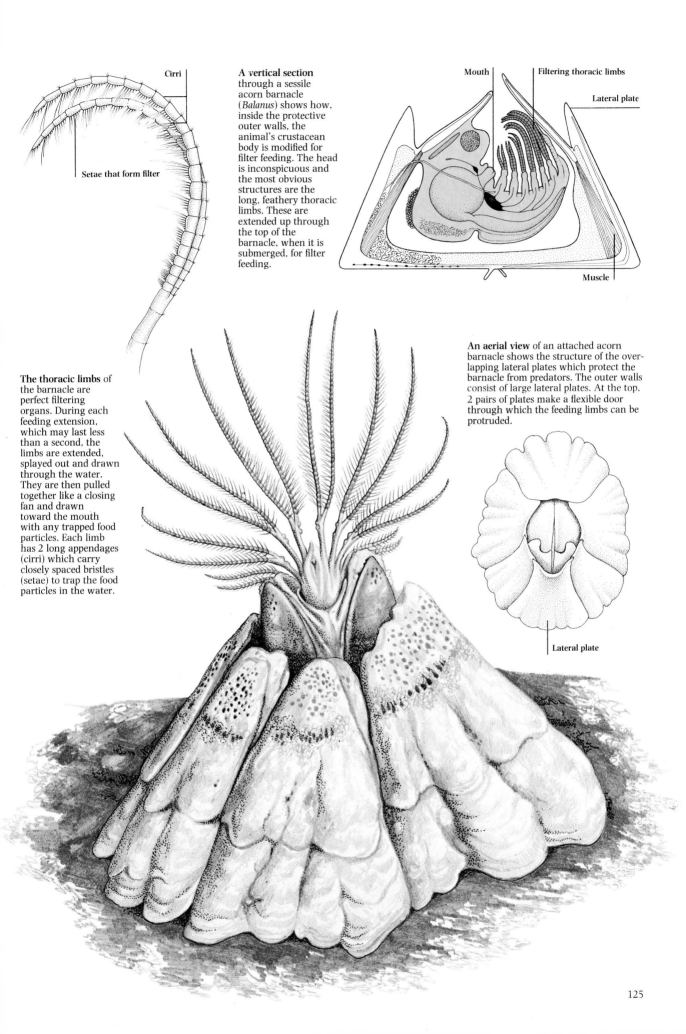

Cirri

Setae that form filter

A vertical section through a sessile acorn barnacle (*Balanus*) shows how, inside the protective outer walls, the animal's crustacean body is modified for filter feeding. The head is inconspicuous and the most obvious structures are the long, feathery thoracic limbs. These are extended up through the top of the barnacle, when it is submerged, for filter feeding.

Mouth

Filtering thoracic limbs

Lateral plate

Muscle

The thoracic limbs of the barnacle are perfect filtering organs. During each feeding extension, which may last less than a second, the limbs are extended, splayed out and drawn through the water. They are then pulled together like a closing fan and drawn toward the mouth with any trapped food particles. Each limb has 2 long appendages (cirri) which carry closely spaced bristles (setae) to trap the food particles in the water.

An aerial view of an attached acorn barnacle shows the structure of the over-lapping lateral plates which protect the barnacle from predators. The outer walls consist of large lateral plates. At the top, 2 pairs of plates make a flexible door through which the feeding limbs can be protruded.

Lateral plate

The Greater Flamingo

With the teetering grace of a stilt walker, the flamingo, a tall, long-legged, long-necked bird, its plumage in part an improbable candy pink or orange colour, wades through the water. As it does so, this striking bird, whose black-tipped wings make it equally impressive in flight, puts its filter feeding mechanism into action. Of all birds flamingoes are most comprehensively adapted for filter feeding—and they are also an ornithological enigma, for their bodies have become so distorted and moulded to accommodate the filter feeding habit that all unambiguous clues to their ancestry have disappeared.

The world probably has only four flamingo species. One of them, the greater flamingo, *Phoenicopterus ruber*, is split into several distinct-looking and geographically separate sub-species found from Asia through Africa and Europe to Chile. The most successful species is the lesser flamingo, *Phoeniconaias minor*, an inhabitant of Africa and Asia, which represents some 75 per cent of the world total of about 6 million flamingoes.

When feeding and breeding all flamingoes are gregarious creatures and many bird watchers dream of seeing the flocks of lesser flamingoes on Lake Nakuru, in Kenya's Rift Valley, which probably number over 2 million individuals. This enormous bulk of animal life is sustained solely by filter feeding and every day each flamingo consumes about 10 per cent of its own body weight in microscopic food organisms such as minute algae and invertebrates.

To achieve this vast straining rate, flamingoes usually have to feed throughout daylight hours. In the tropics and subtropics, this food straining invariably takes place in brackish or salt water lagoons or lakes. In many parts of the globe these conditions are encountered at high altitudes and one Andean flamingo lives in lakes more than 13,000 feet (3962 metres) above sea level.

The flamingo's unique filter feeding system is housed in its huge beak which, in most species, is bent downward at 45 degrees about halfway along its length. The lower half of the beak is much larger than the upper and has a deep, trough-shaped depression which is capped by the lid-like upper bill when the beak is closed. As the bird wades, the head at the end of its sinuous neck is held so that the front half of the beak lies horizontally below the water surface; the beak is upside down so that the upper bill is under the lower bill. In this position the special filtering structures on the upper beak can, with the aid of the tongue, sieve out food particles.

Whale birds or prions are a group of 14 species of petrel from the waters around the Antarctic. They are the flamingo's only real competitors for the title of top bird filter feeders. They feed on animal plankton by straining sea water through the lamellae which fringe the upper bill. Like the rorqual whales, they force the water through the sieve with an enlarged tongue and trap the tiny organisms.

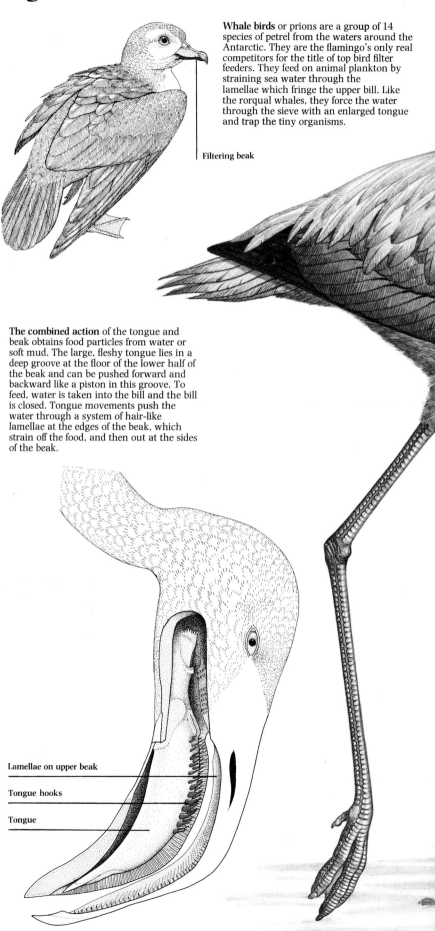

Filtering beak

The combined action of the tongue and beak obtains food particles from water or soft mud. The large, fleshy tongue lies in a deep groove at the floor of the lower half of the beak and can be pushed forward and backward like a piston in this groove. To feed, water is taken into the bill and the bill is closed. Tongue movements push the water through a system of hair-like lamellae at the edges of the beak, which strain off the food, and then out at the sides of the beak.

Lamellae on upper beak

Tongue hooks

Tongue

The greater flamingo, *Phoenicopterus ruber*, is the largest of the flamingo species and stands up to 6 ft (1.8 m) high. It is the best example of a flamingo with a shallow-keeled beak apparatus for feeding. The American flamingo is a sub-species of the greater flamingo.

The hair-like lamellae, the basis of the flamingo's filtering mechanism, are arranged in precise, oblique rows inside the beak so that when the 2 halves of the beak are brought together, the lamellae interlock exactly. They vary greatly in shape, size and spacing according to the part of the beak they are in and the species of flamingo. The outer lamellae, at the margins of the upper jaw, are large, projecting hooks. The inner lamellae are smaller and more closely spaced. Examples of both types are shown.

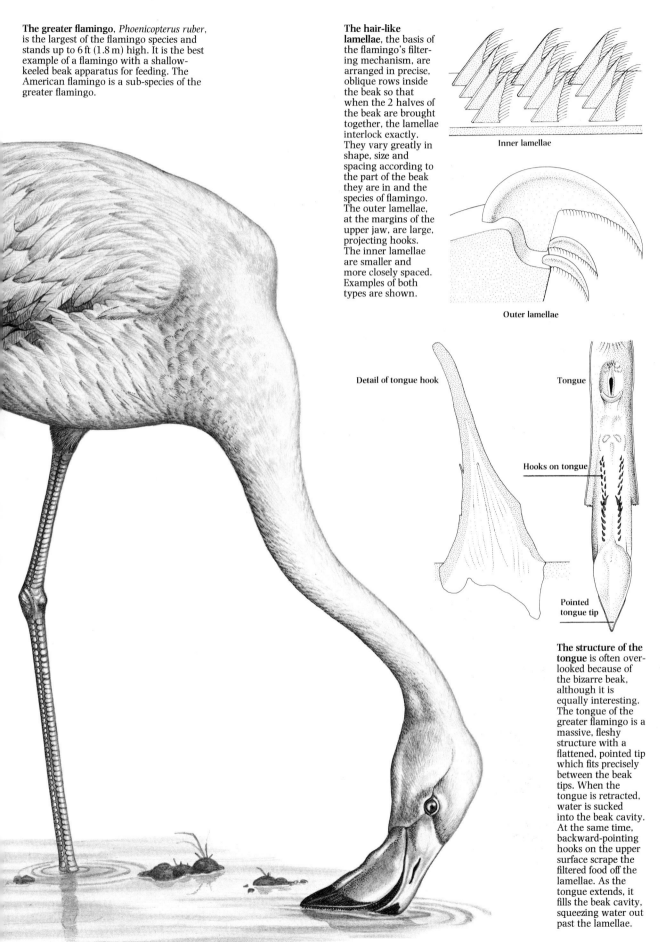

Inner lamellae

Outer lamellae

Detail of tongue hook

Tongue

Hooks on tongue

Pointed tongue tip

The structure of the tongue is often over-looked because of the bizarre beak, although it is equally interesting. The tongue of the greater flamingo is a massive, fleshy structure with a flattened, pointed tip which fits precisely between the beak tips. When the tongue is retracted, water is sucked into the beak cavity. At the same time, backward-pointing hooks on the upper surface scrape the filtered food off the lamellae. As the tongue extends, it fills the beak cavity, squeezing water out past the lamellae.

127

PARASITES AND PARASITOIDS: LIVING-IN PREDATORS

In everyday language, a parasite is someone who sponges on others and takes what is rightfully theirs without labouring fairly for the rewards. This implied and emotive criticism is difficult to ignore when discussing animal parasites; but despite the damage, disease or even death caused by many parasites to their hosts—the organisms at whose expense they make a living—parasites are best regarded as creatures with specialized feeding methods rather than as pests.

Parasites may be animals, plants or micro-organisms, but all are committed to living in close association with their hosts. The relationship between host and parasite is often highly specific from the parasite's viewpoint: for any particular parasite only a narrow range of species will be adequate and appropriate hosts. Living in or on the host for at least part of its life span, the parasite nearly always depends on the host for its food. Almost any type of host tissue can be removed to provide sustenance for the parasite. Lampreys, for example, live on blood, lice on skin, and tapeworms on the digested food in the gut.

The eventual outcome of this food removal, whatever its pattern, always results in an unbalanced distribution of harm and benefit in the host-parasite relationship: the parasite gains and the host suffers. Parasites may rob the host of vital organs or of renewable ones such as blood or skin so quickly that the host debilitates itself in trying to replace the lost tissues. Equally, the parasite may produce toxins that result in direct injury to the host, or so alter its behaviour that it is more likely to be killed by an outside predator. But whatever specific damage or pathological changes actually occur, organisms that play host to parasites usually die more quickly and reproduce more slowly than those that do not have parasites.

The hosts of many parasites can be considered as their habitats—albeit ones that are sparsely and patchily distributed in the parasite's actual environment. Parasites are thus faced with the problem of getting from one host individual to another and large numbers of parasites die in the harsh outside world while searching for the comparatively cosy conditions that can sustain them within the correct hosts. The very damage that parasites cause can also be their undoing. A host attacked by a large number of parasites runs a high risk of an early death. If it dies, a heavily infected host behaves involuntarily like a kamikaze pilot, for parasites rarely survive the death of their hosts.

Parasites have met this two-pronged problem by evolving a range of adaptations to offset their difficulties. Most striking is the immense reproductive capacity common to almost all parasites. Parasites often reproduce hundreds or thousands of times faster than their hosts, thus enabling their populations to carry on despite high mortality rates.

An interesting halfway house between host-parasite associations and predator-prey interactions is occupied by creatures called parasitoids. Most, such as the ichneumon flies, are insects that live normal herbivorous or carnivorous adult lives. The female lays her eggs on or in a host. If not already inside the host, the parasitoid larvae that hatch from the eggs burrow in, and gradually consume their hosts from the inside.

The Lamprey

The parasitic way of life **might** seem beyond the scope of vertebrates, but these most complex of creatures do make use of parasitism in a number of ways. The cuckoos, for instance, are the best-known examples of brood parasites—birds that lay their eggs in the nests of other bird species which then act as unknowing foster parents. Living or feeding on the surfaces of their hosts as ectoparasites, several sorts of vertebrates gain sustenance by eating blood. The vampire bats live in this way, as do the lampreys, creatures that are probably the most specialized underwater blood feeders.

The lampreys, and their close relatives the hagfish, are slimy-skinned, eel-like animals that belong to the most primitive group of vertebrates alive today. Strictly, lampreys and hagfish are not fish at all, for in place of a true vertebral column their bodies are supported by an

elastic rod or notochord—the primitive forerunner of the backbone—studded with lumps of cartilage. And unlike other fish, neither lampreys nor hagfish have true jaws.

Some 30 species of lampreys are known around the world in both northern and southern temperate zones, and all the adults are blood-feeding ectoparasites. They attach themselves to the sides of fish with a sucking mouth disk armed with rows of rasping teeth that pierce into the blood system.

All lampreys have complex life cycles involving two distinct feeding methods. Immature larval lampreys live buried in the mud of freshwater streams and ditches where they exist as filter feeders. When they mature, these larvae go to sea and spend a considerable time as blood-feeding ectoparasites before returning to freshwater spawning grounds to breed.

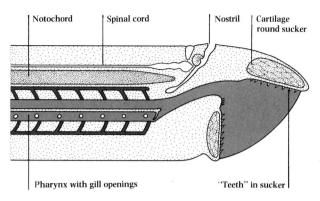

Notochord · Spinal cord · Nostril · Cartilage round sucker

Pharynx with gill openings · "Teeth" in sucker

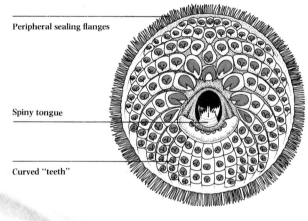

Peripheral sealing flanges

Spiny tongue

Curved "teeth"

Complex muscles operate the mouth sucker and enable the lamprey's tongue to be protruded and moved to rasp at the host's skin. The mouth opens into a cavity that divides into a dorsal oeso-phagus and a tube which runs to the 7 paired gill pouches. When feeding, the lamprey pumps respiratory water in and out of the gills using the muscles surrounding them.

The mouth sucker is fringed with a series of sealing flanges or lips, which enables it to grip the skin of a host fish. On the roof of the sucker are curving rows of pointed teeth— horny thickenings of skin. In the centre of the sucker is the mouth cavity itself, through which pro-trudes a toothed tongue. This is used as a rasp to scrape away the host's skin before feeding.

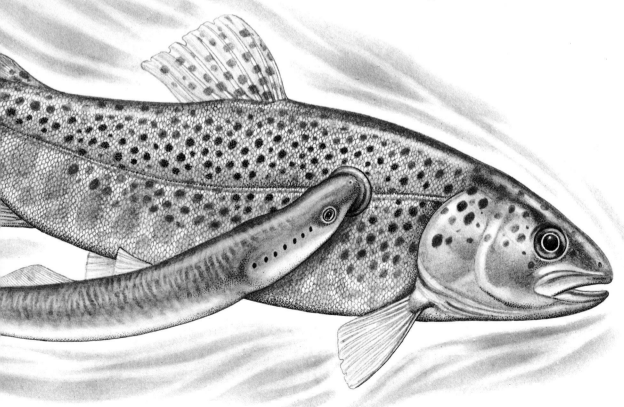

The river lamprey, *Lampetra fluviatilis*, is about 20 in (51 cm) when fully grown. River lampreys are migratory fish that spend at least a year feeding in the sea, as external parasites, on fish blood. The typical fish that they feed on are migratory, brackish-water species such as sea trout and shad. After fattening up on marine fish, the lamprey returns to the rivers in autumn and spends the winter in fresh water usually without feeding. Spawning occurs in the spring and afterwards the lamprey dies. The adult is dull grey in colour, lighter on the ventral side. Fish that have just returned to the river to spawn are golden.

The Louse and the Tapeworm

The human body louse and the tapeworm are parasites committed to perpetual residence on or in their hosts. Although such parasites may seem insignificant, over 10,000 body lice have been found on a single item of clothing and the fish and beef tapeworms, which infect humans, can be over 60 feet (18.3 metres long).

Like its free-living, non-parasitic insect ancestors, the louse has a small body divided into a head, thorax and abdomen. Although it has no wings, the body bears three pairs of legs, plus paired eyes and antennae. All lice belong to one of two quite distinct groups—the sucking lice or Anoplura, which live on mammals and include the body louse, and the chewing or feather lice of the sub-order Mallophaga whose hosts are almost always birds.

When feeding on man, lice do not rob him of damaging amounts of blood, but they can be disastrously efficient at transmitting diseases—particularly epidemic typhus and relapsing fever—caused by micro-organisms for which lice act as vectors.

Supremely adapted creatures, tapeworms have undergone drastic evolutionary changes to their body plans in becoming internal parasites. They are dependent, as adults, on the cavity within the vertebrate gut for food and shelter. Tapeworms have lost the intestinal organs possessed by the non-parasitic, flatworm ancestors and, instead, absorb ready-digested organic molecules through their body walls.

All true tapeworms are constructed to the same basic blueprint. At the "head" end is an attachment organ, the scolex, containing many sense cells and what passes for a brain. This scolex anchors the worm to the lining of its host's gut and stops it from being swept backward by the rush of passing food and by the waves of muscle contraction that help push the food along.

Immediately behind the tapeworm scolex is a rapid-growth zone, "the neck", which produces a series of buds or segments. These are pushed toward the tapeworm's tail end by the production of yet younger segments to form a chain of segments or proglottids that may contain many thousand links and measure several yards in length.

As it matures, each segment develops reproductively as if it were an entire worm, producing first male then female sex organs. Copulation takes place between segments so that fertilized segments become packed with eggs. These segments can either release their eggs into the host gut or break off from the end of the tapeworm as living egg packets, but in both cases the eggs hatch to release larvae that start the next phase of a complex life cycle that may involve at least two types of host.

A marine tapeworm larva may find its way into the tissue of an oyster where it causes the irritation which eventually produces a pearl.

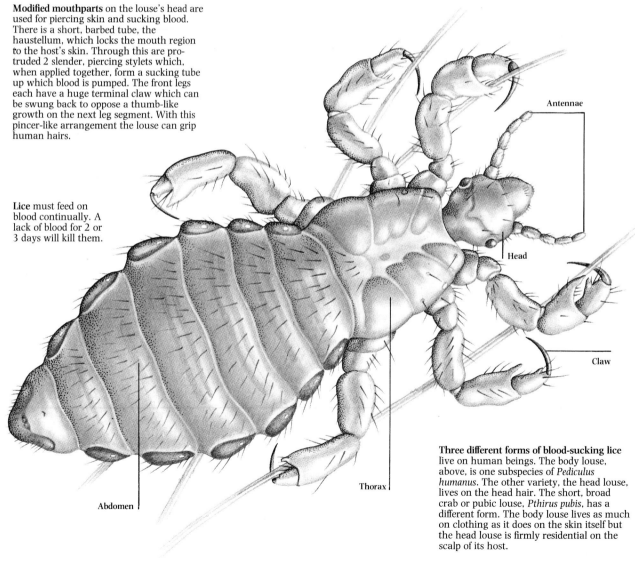

Modified mouthparts on the louse's head are used for piercing skin and sucking blood. There is a short, barbed tube, the haustellum, which locks the mouth region to the host's skin. Through this are protruded 2 slender, piercing stylets which, when applied together, form a sucking tube up which blood is pumped. The front legs each have a huge terminal claw which can be swung back to oppose a thumb-like growth on the next leg segment. With this pincer-like arrangement the louse can grip human hairs.

Lice must feed on blood continually. A lack of blood for 2 or 3 days will kill them.

Antennae

Head

Claw

Thorax

Abdomen

Three different forms of blood-sucking lice live on human beings. The body louse, above, is one subspecies of *Pediculus humanus*. The other variety, the head louse, lives on the head hair. The short, broad crab or pubic louse, *Pthirus pubis*, has a different form. The body louse lives as much on clothing as it does on the skin itself but the head louse is firmly residential on the scalp of its host.

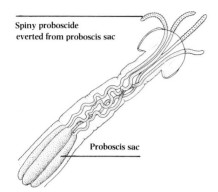

Spiny proboscide everted from proboscis sac

Proboscis sac

Hooks, suckers and muscular clamps on the tapeworm's head provide the basic repertoire of anchoring devices, but in number, shape and combination each species has its own particular design. The tetrarhynch, above, has 4 long, eversible organs each covered with hundreds of backward-facing barbs.

The muscular organ of *Acanthobothrium* consists of a ring of 2-pronged, backward-pointing hooks. Beneath each pair of hooks is a lobed, muscular sucker. The combined activities of hooks and sucker mean that once attached to the host's gut wall, the tapeworm is difficult to remove.

The outer surface of the tapeworm has a complex structure only recently revealed with the electron microscope. The surface is thrown into myriads of thin, living projections which expand the surface area for food absorption. Each projection has a backward-pointing spine at its tip which perhaps helps to prevent the tapeworm being pushed backward in the gut. Mucus materials are secreted in the surface and protect the exposed covering from digestive attack.

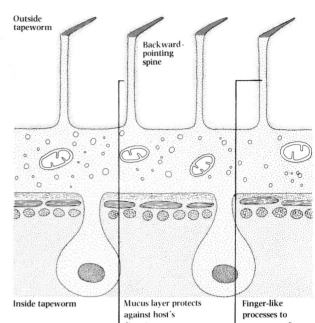

Outside tapeworm

Backward-pointing spine

Inside tapeworm

Mucus layer protects against host's digestive enzymes

Finger-like processes to increase surface area for food absorption

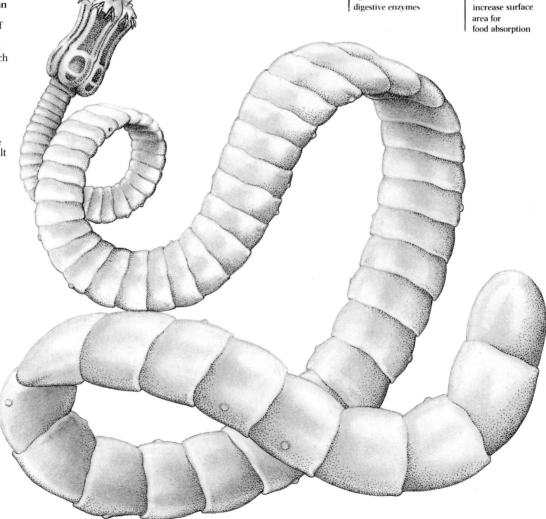

Tapeworms are specialized animals endowed with sense organs and a muscular system which enables them to move about actively in the gut. They often inhabit a limited zone of gut and if moved will unerringly return to their proper microhabitat. Their surface is designed for the absorption of dissolved food materials and also for active protection against the immediate external environment which contains unpleasant substances like bile salts and digestive enzymes. The tapeworm above, from the genus *Acanthobothrium*, is restricted to living in elasomobranch fish such as sharks and rays.

The Ichneumon Fly

Slowly and from within, the larvae of parasitoid insects eat their hosts. These larvae usually attack invertebrates—often other insects—and nearly always kill the animals in which they live and feed. Death results either from the destruction of crucial internal organs or by mortal wounds inflicted as the mature larvae cut or chew their way out of their hosts.

The almost inevitable death of the parasitoid's host clearly separates its life-style from that of a parasite, for even when invaded by parasites, most hosts survive. Of the groups of insects that have parasitoid members some do, in fact, approach orthodox parasitism. Parasitoid wasps, for example, which use scale insects as hosts, may extend the relationship long enough for the host to lay eggs. And some ichneumon species have larvae that can complete their development and emerge without killing the insects which have involuntarily provided food from their own internal organs.

Among the world's 29 insect orders there are two that show a wide diversity of parasitoid life-styles—the Diptera or two-winged flies, and the Hymenoptera, the group to which ants, bees and wasps belong. The dipterans that live as parasitoids do not necessarily use insects as their hosts. Many, if not all, of the marsh flies of the family Sciomyzidae, for instance, have larvae that are specialized parasitoids of land and amphibious snails.

The female of one marsh fly, *Pteromicra*, lays her eggs on the shells of watersnails of the genus *Lymnaea* found on floating vegetation near pond margins. Within each snail, an active, ravenous larva hatches, bores into the soft parts of the snail and spends about six days feeding on the snail's tissues. By the time the larva has completed its development the snail is dead and the larva pupates in the now empty shell.

The Hymenoptera, a varied and successful insect order, comprises some 200,000 species, a total second only to the beetles. Within the group, apart from the plant-eating sawflies and socialized ants, wasps and bees, the majority are parasitoids whose larvae develop in other insects. Ichneumons, for example, are slender-bodied hymenopterans classified in the superfamily Ichneumonoidea and all are parasitoids. Ichneumon abdomens are usually long and thin and the female is often equipped with a long ovipositor from which eggs are ejected. This structure is especially conspicuous and important in forms that have to bore through layers of wood to reach host insects that live or feed in wood.

A parasitoid usually lays only one egg in each host. If more were laid there would not be enough food for the hatching offspring. Another parasitoid hymenopteran, *Trissolcus basalis*, which lays its eggs in the eggs of shield bugs, "marks" each egg that has been parasitized. The female moves the tip of the abdomen in a figure-eight over the egg, leaving physical or chemical traces to warn off other potential parasitoids.

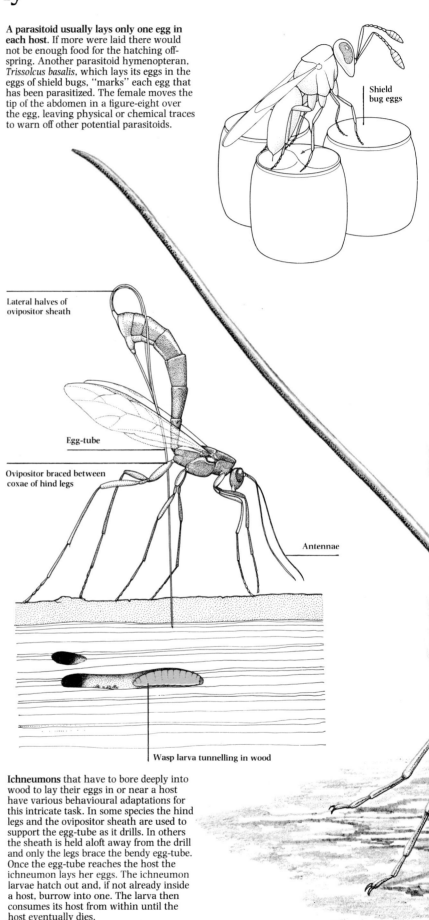

Shield bug eggs

Lateral halves of ovipositor sheath

Egg-tube

Ovipositor braced between coxae of hind legs

Antennae

Wasp larva tunnelling in wood

Ichneumons that have to bore deeply into wood to lay their eggs in or near a host have various behavioural adaptations for this intricate task. In some species the hind legs and the ovipositor sheath are used to support the egg-tube as it drills. In others the sheath is held aloft away from the drill and only the legs brace the bendy egg-tube. Once the egg-tube reaches the host the ichneumon lays her eggs. The ichneumon larvae hatch out and, if not already inside a host, burrow into one. The larva then consumes its host from within until the host eventually dies.

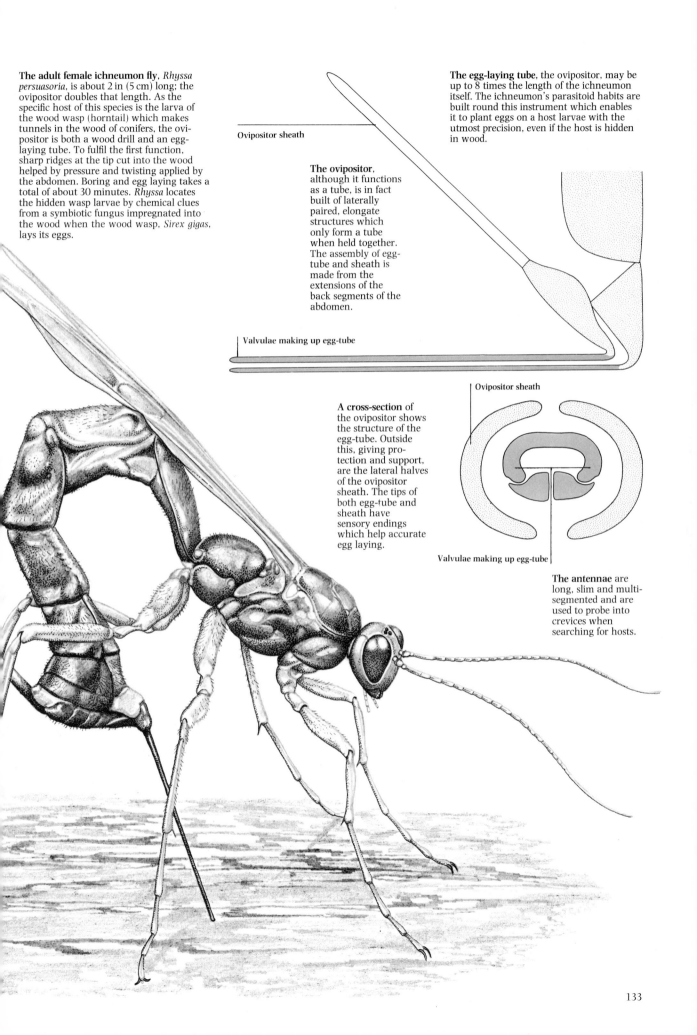

The adult female ichneumon fly, *Rhyssa persuasoria*, is about 2 in (5 cm) long; the ovipositor doubles that length. As the specific host of this species is the larva of the wood wasp (horntail) which makes tunnels in the wood of conifers, the ovipositor is both a wood drill and an egg-laying tube. To fulfil the first function, sharp ridges at the tip cut into the wood helped by pressure and twisting applied by the abdomen. Boring and egg laying takes a total of about 30 minutes. *Rhyssa* locates the hidden wasp larvae by chemical clues from a symbiotic fungus impregnated into the wood when the wood wasp, *Sirex gigas*, lays its eggs.

Ovipositor sheath

The egg-laying tube, the ovipositor, may be up to 8 times the length of the ichneumon itself. The ichneumon's parasitoid habits are built round this instrument which enables it to plant eggs on a host larvae with the utmost precision, even if the host is hidden in wood.

The ovipositor, although it functions as a tube, is in fact built of laterally paired, elongate structures which only form a tube when held together. The assembly of egg-tube and sheath is made from the extensions of the back segments of the abdomen.

Valvulae making up egg-tube

A cross-section of the ovipositor shows the structure of the egg-tube. Outside this, giving protection and support, are the lateral halves of the ovipositor sheath. The tips of both egg-tube and sheath have sensory endings which help accurate egg laying.

Ovipositor sheath

Valvulae making up egg-tube

The antennae are long, slim and multi-segmented and are used to probe into crevices when searching for hosts.

133

SCAVENGERS: THE WASTE DISPOSERS

Drawn up in the most simple way, the food chains and webs that link the lives of the world's natural organisms are like hypothetical one-way road systems. According to this restricted scheme of things, energy is trapped and used by growing plants, plants are eaten by herbivores and herbivores are food for low-level carnivores which, in turn, fall prey to the top predators. But, just as on most real highways, traffic moves in both directions, so energy flows in two ways up and down the food chains, for not all—nor even the majority—of animals, plants and micro-organisms are eaten alive. Many die through causes other than predation, such as disease or old age, and even predators leave remains from their kills. This great bulk of dead organic material is the fuel source that ensures the two-way energy flow and is a vast biological resource that has not been allowed to go untapped. The living world has no room for waste and if a resource exists that can be used to advantage, some creature will adapt its life to fulfil such a role.

The armies of organisms that have evolved to act as natural refuse processors are called decomposers if they are small and scavengers if they are large. Microscopic decomposers make up the vast proportion of the first group. Bacteria, fungi and single-celled animals or protozoans are all decomposers, breaking down complex molecules into simple ones. These simple molecules may be used for growth and are thus returned to the living part of the ecosystem. Wherever animals or plants die the decomposers appear *en masse*, whether in leaf litter, in the soil, on the bodies of dead animals or in the sediments on the ocean bottom. These minute organisms are at the heart of the natural recycling processes which ensure that organic nutrients and vital minerals continue to circulate among the living creatures of any ecosystem.

It is to these bacteria, fungi and protozoans that industrial man turns when he needs to break down organic wastes. One of the most common modern methods of sewage treatment is the "activated sludge" process in which sewage is mixed and oxygenated while natural decomposers eat their way through the dead organic wastes.

Because of their size, scavengers such as burying beetles are easier to observe and to understand. Most detritus-feeding invertebrates, including burrowing aquatic worms, are probably both predators and scavengers that will remove both small living organisms and dead material from detritus. Among the larger vertebrate scavengers is a whole spectrum of life-styles ranging from almost pure scavengers such as vultures to specialist predators that will sometimes help themselves to carrion if it becomes available. Between these extremes are vertebrates that subsist largely on dead animal material but retain the capacity for active predation when the opportunity arises. Among the different members of the African hyena family, for example, striped hyenas are almost entirely scavenging in their habits, spotted hyenas are mixed predators and scavengers while aardwolves are predatory insectivores that specialize in finding and eating underground concentrations of termites but may scavenge for food on occasion.

The Burying Beetle

If success can be judged by sheer numbers, then the beetles, insects of the order Coleoptera, must top the animal list, for more than 350,000 beetle species exist today, and of all the world's animals, one in every three is a beetle.

Ranging in size from minute weevils less than $\frac{1}{16}$ inch (1.5 millimetres) long to the giant 6-inch (15 centimetres) Hercules beetles of tropical America, beetles have become adapted to a wide range of life-styles. Many are active predators but large numbers are committed to lives as carrion eaters and scavengers.

Using their strong legs and blunt heads, the burying or sexton beetles shift soil from beneath animal carcasses so that they sink into the ground. Once interred the corpse becomes a perfect site in which eggs can be laid. When the larvae hatch from these eggs, they have a ready and ample food supply. Many adult beetles also live on carrion. The carrion beetles, equipped with powerful jaws and mobile heads, are among the first scavengers to go to work on dead animals. While their jaws "worry" the carrion, the spined legs are pushed against it to give extra purchase as the beetle wrestles with tough tissues. The active rove beetles search out dead bodies of small animals—they too have powerful jaws with which to tackle their "prey".

Whatever their life-style, all beetles share many characteristics, most notably the arrangement of the two pairs of wings. The front pair have been converted from active aerofoils into a protective case to house the functional, membranous hind wings. In the closed position the wing cases or elytra are brought together to meet over the beetle's abdomen and the hind wings are folded, either transversely or longitudinally, to fit under them.

The water scavengers of the beetle family, Hydrophilidae, are shiny black, aquatic insects that derive most of their food from dead and decaying organic material. The species below is about 1½ in (4 cm) long and swims well with alternate kicks of its legs. On its undersurface is a pointed spine which can be used for defence against predators. Fish, ducks or even a fisherman can receive a painful jab if they happen to touch this beetle.

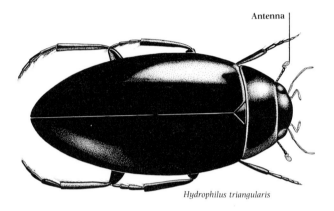

Antenna

Hydrophilus triangularis

THE HORN MOTH
Even horns—the final remnants of a kill—can be used. The horn moth, *Tinea deperdella*, related to the clothes moth, lays its eggs on the horns. The horn moth caterpillars, which can digest keratin, feed on the horns.

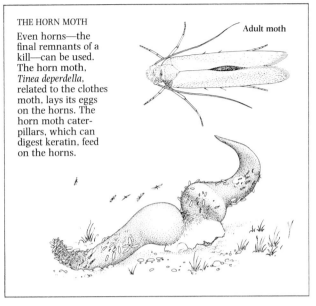

Adult moth

The burying beetle, *Necrophorus mortuorum*, gets its common name from the fact that in groups, it will sometimes bury a small corpse underground before starting feeding and egg-laying. This habit however is not essential to the beetle's life-style. An adult burying beetle is about ½ in (12 mm) long with clubbed tentacles that probably carry chemical sense organs to help it track down rotting flesh.

The warning coloration of the burying beetle, striking black and ochre bands, probably refers to its unpleasant musky smell and the evil-smelling fluid it can exude from the mouth if attacked.

A group of burying beetles have discovered the dead body of a sparrow. They will feed on the flesh of the carcass and the females will then lay their eggs in it. The larvae which hatch from the eggs are also carrion-eaters and have cylindrical, fleshy bodies and weak legs.

135

Hyenas and Jackals

Hyenas and jackals are usually thought of as slinky, rather cowardly animals that steal a share of the kills made by lions or cheetahs. Consequently, it is often believed that they are on the sidelines of the principal predator-prey interactions that occur in the African plains.

They are indeed scavengers, but the same description can be applied to almost any carnivore. Virtually all hunting animals prefer an easy meal to one that involves a chase or an active kill. Hyenas certainly will make use of already dead food whenever they can, but the detailed observations of Hans Kruuk in the Serengeti and in the Ngorongoro crater of Tanzania have shown that they are also efficient predators. Kruuk, studying the spotted hyena, *Crocuta crocuta*, of the East African plains, found that in the Ngorongoro hyenas killed 93 per cent of their food and in the Serengeti 68 per cent. Thus only a small percentage of their diet consists of material scavenged

from the kills or the natural deaths of other animals.

How then has the hyena earned its reputation as the archetypical scavenger of carrion? The reasons are complex. First, the spotted hyena has probably been mistaken for its near relative, the striped hyena, *Hyaena hyaena*, which is a more thorough-going scavenger, getting the bulk of its food from the refuse around human habitations in India, the Middle East and North Africa. Second, the main killing time for most hyena groups is at night, when game hunters, game wardens and tourists are not likely to be making detailed field observations. Third, much of the competition that occurs between lions and hyenas has been shown by Kruuk to have been misinterpreted in the past. When lions and hyenas are seen in the daytime contesting over a dead animal, the hyenas are just as likely to have killed it as the lions are.

There are two common species of jackals in the Serengeti—the golden,

Canis aureus, which is found primarily in open grassy plains, and the black-backed, *Canis mesomelas*, which is seen more regularly in bush country. The two species often scavenge together at a kill but, like the larger hyenas, can also make predatory kills of their own.

In East Africa jackals eat small mammals like young gazelles, insects, fruit and meat and bones scavenged from the kills of larger predators. As two common scavenging animals in the same ecosystem, jackals and hyenas are frequently observed together at the same kill. Hyenas will tolerate jackals at a kill. When feeding on lion kills, however, jackals are able to take more food than hyenas because their acceleration away from a lion is much faster than the hyenas'. In addition, it is possible that by their efficient detection and killing of young gazelles, jackals provide hyenas with an extra food source to scavenge. To a minor degree then, the predatory and scavenging interactions between these animals are mutually beneficial.

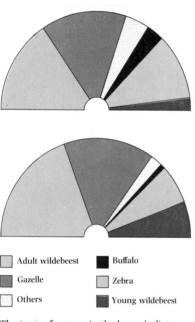

The types of carcass in the hyena's diet are similar whether only prey killed by the hyena is considered or scavenged prey is included. The prey is remarkably similar to that of the lion and in this aspect the animals show a great deal of overlap.

Adult wildebeest
Gazelle
Others
Buffalo
Zebra
Young wildebeest

A solitary black-backed jackal and 3 spotted hyenas feed on an old giraffe kill. Hyenas are perfectly designed to make use of almost every fragment of a carcass except horns and teeth; everything else, even bones, disappear into that most murderous of animated waste-disposal units, the hyena's mouth. A group of 25 hyenas was observed to devour 3 adult zebras and a foal in just over 30 minutes.

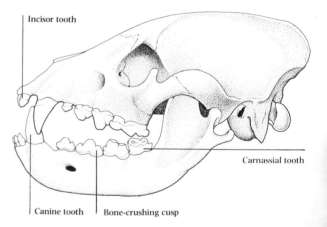

The hyena's skull reveals its effective tooth armament. The canine teeth, $1\frac{1}{2}$ in (4 cm) long, are attack weapons and the incisors are used for stripping soft flesh from a carcass. The conical, bone-crushing molars can demolish almost any bones to powdery chips. Behind them is a strongly bladed carnassial for stripping gristle and cutting skin as well as gnawing flesh from bones.

Incisor tooth

Carnassial tooth

Canine tooth   Bone-crushing cusp

The droppings of the hyena are white, not, as was believed by the Masai Africans, because hyenas eat ashes, but because they contain a high percentage of bone powder.

A JACKAL KILL

The young of small gazelles are the largest prey jackals can kill. A pair of jackals attack the mother and young. One jackal then chases and distracts the mother while the other kills the fawn. Both then eat the prey. Jackals hunting in this way catch two-thirds of the animals they chase.

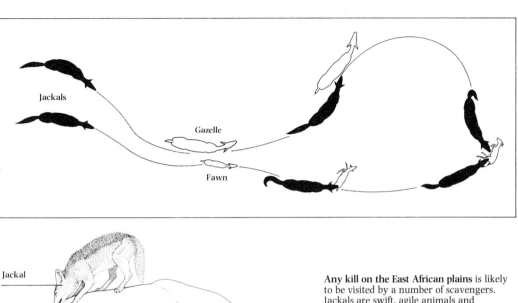

Jackals

Gazelle

Fawn

Jackal

Hyena

Any kill on the East African plains is likely to be visited by a number of scavengers. Jackals are swift, agile animals and dash in among the others to grab a fragment of food and dash off again. Many of the kills they feed on are hyena kills and, when the carcass is large, hyenas and jackals can coexist. A hyena is quite capable of eating its way right into the body cavity with its massive teeth, while the jackals take fragments from the surface.

The White-backed Vulture

Vultures are scavenging birds similar to, but quite distinct from, the falcons. They are split into two separate families: the Old World species, typified by the vultures of Africa and Asia, and the New World species, such as the turkey vulture and the magnificent condors of North and South America.

Of the 20 existing vulture species, all except one are specialist carrion-eaters. The odd bird out is the palm-nut vulture, *Gypohierax angolensis*, which feeds mainly on the fruit of the oil palm. Like its more typical relatives, however, it also eats fish and animal remains.

Almost all vultures are exceptionally large birds. The biggest African species, such as the king vulture, *Sarcogyps calvus*, weigh between 10 and 15 pounds (4.5 to 6.8 kilogrammes) and have a wingspan of up to 9 feet (2.7 metres). Both the Andean condor,

Vultur gryphus, and the exceedingly rare Californian condor, *Gymnogyps californianus*, have a wingspan of almost 10 feet (3 metres).

A number of structural adaptations enable vultures to locate and feed on carrion. Their powerful, hooked bills are capable of cutting into the skin and flesh of large carcasses. Their feet, relatively feeble compared to those of other birds of prey, are designed primarily for running and walking rather than for grasping prey. Their immense, broad wings have wide-spaced primary feathers that provide an extremely efficient high-lift soaring and gliding aerofoil. As a consequence, vultures are able to remain aloft on tropical thermal updraughts for long periods without active wing beats. In most species, part of the head and neck is bare of feathers except for a thin covering of down.

Old World vultures inhabit southern Europe, Africa and southern Asia. Scarcely ever seen in heavily wooded areas or swamplands, they are commonly found in flat plains or mountainous terrain, where their soaring method of locating food is apt to meet with the most success. In general, New World vultures enjoy a wider range of habitats; condors, however, confine themselves to mountainous regions.

Although vultures are specialized for a scavenging existence, some have retained the capacity to make kills of their own. On the Peruvian guano islands, for example, the turkey vulture, *Cathartes aura*, also called the turkey buzzard and John Crow, takes eggs and young from sea bird nests. Likewise, a lappet-faced vulture in Africa will occasionally kill a small gazelle.

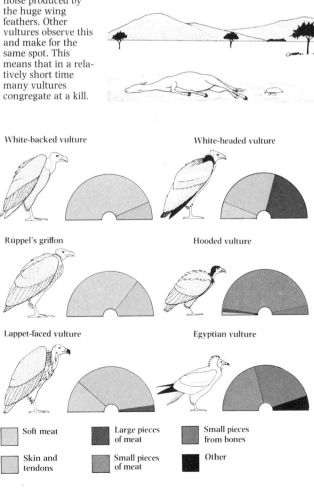

When searching for food, each vulture soars high in the sky patrolling a roughly defined area which does not greatly overlap the areas of others. If one vulture sights carrion, it plummets down with a whistling noise produced by the huge wing feathers. Other vultures observe this and make for the same spot. This means that in a relatively short time many vultures congregate at a kill.

White-backed vulture

White-headed vulture

Rüppel's griffon

Hooded vulture

Lappet-faced vulture

Egyptian vulture

☐ Soft meat

☐ Skin and tendons

☐ Large pieces of meat

☐ Small pieces of meat

☐ Small pieces from bones

☐ Other

Each vulture species has its own characteristic pattern of removing different types of food material from the carcass and studies reveal these differences. Thus the apparently undignified scramble for food is not quite as it seems and competition between the species is lessened. The order in which species visit a carcass also has a pattern to it. White-headed vultures, although relatively rare—3 per cent of the vulture population—are often first. The large species are the only ones capable of cutting into the carcass; smaller birds subsist on the scraps of the kill.

Both Old and New World vultures are almost unique among birds in having bald heads. This feature seems to be connected with the carrion-eating life-style. Feather maintenance by cleaning and preening is an important part of a bird's life. The vulture's feeding method, plunging head and neck deep into a decomposing carcass, would make cleaning a long and difficult process, hence the elimination of the head feathers. The theory is supported by the fact that forms such as the white-headed vulture, which plunges into carcasses, have bald heads and necks, while the lammergeier, which feeds on bones, has a fully feathered head.

White-headed vulture

Lammergeier

The white-backed vulture, *Pseudogyps africanus*, is the most common of the 6 species usually seen in the East African plains. In fact, 64 per cent of vultures in the Serengeti are this species. When feeding on a carcass, it appears to specialize on the soft, easily removed meat. It roosts colonially in trees, making huge nests lined with fresh green leaves. Each pair of vultures produces a single egg. The female and the young vulture are fed with carrion regurgitated by the male bird.

Carnivorous Plants

The few carnivorous plants scattered around the world are both entrancing and scientifically perplexing. Green plants are supposed to be autotrophic, that is, able to sustain themselves on carbon dioxide, water and mineral salts with the help of sunlight energy and chlorophyll. It was therefore quite a shock for early investigators to discover that plants could catch living animals as well as digest and assimilate them.

This remarkable phenomenon—that of plants eating insects, protozoans, crustaceans and even mice—is now a fully authenticated fact. Some 367 separate species of green plant from five different families are known to obtain food in this way. Such activity can be understood in the context of the wide range of strategies that plants have adopted in order to exist in soil or water conditions that are deficient in particular types of mineral salts. Many plants, for example, enter into a symbiotic partnership with mycorrhizal fungi that take

over the outer layers of their roots and stretch out into the soil. These fungal mats suck scarce, yet crucial, phosphates out of the ground for the plants more efficiently than do the plants' own root hairs. Similarly, a number of plants such as floating tropical ferns (*Azolla*) and important crops such as lucerne, soya beans and peas have symbiotic, nitrogen-fixing bacteria living inside their own cells. In habitats with low nitrate levels, these invaluable partners pull atmospheric nitrogen out of the air and combine it into organic molecules that the plant can use.

Carnivorous plants use a less friendly mineral-gaining strategy. Almost all live in nitrogen-poor soils or water systems and, instead of using symbiosis to meet their nitrogen shortfall, they resort to killing animals and then digesting nitrogen-rich components such as amino acids from them.

The majority of carnivorous plant species use modified leaves to catch

animals. Their scope of trapping techniques, though, is quite wide. Basically, there are two active methods and three distinct passive techniques. The most famous is the active "gin" or snaptrap method exhibited by the Venus's flytrap, *Dionaea muscipula*, as well as by the five aquatic carnivorous plants in the genus *Aldrovanda*. The only other active method is that used by the bladderworts, genus *Utricularia*. The bladders that give the group its name serve as active suction traps to draw in water and prey. Bladderworts represent the most successful plant experiment in carnivorous living—more than 150 species are found throughout the world.

The three passive catching techniques are known as pitfall (pitcher) traps, adhesive droplet (flypaper) traps and "lobster pots". Insects and other prey fall into the first type of trap, stick to the second and move into but cannot escape from the third.

The one species of Venus's flytrap, *Dionaea muscipula*, is restricted to eastern North America. First discovered in 1759, it perplexed early naturalists until they realized that the leaves were constructed like traps to catch flies. Charles Darwin, in his book *Insectivorous Plants*, called *Dionaea* "the most wonderful plant in the world".

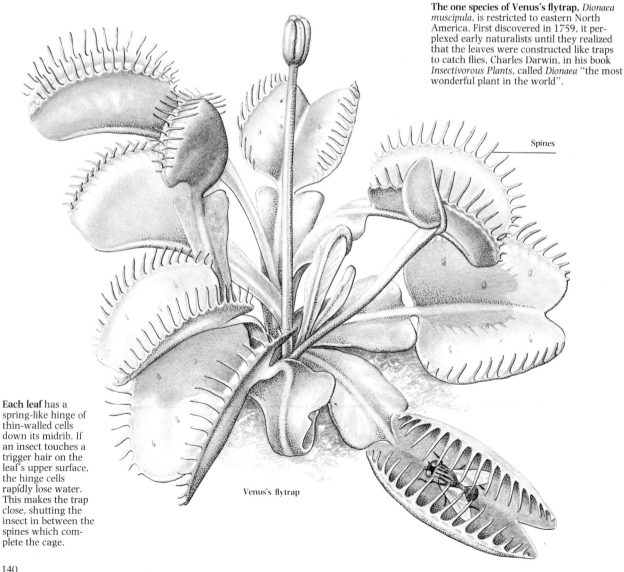

Spines

Each leaf has a spring-like hinge of thin-walled cells down its midrib. If an insect touches a trigger hair on the leaf's upper surface, the hinge cells rapidly lose water. This makes the trap close, shutting the insect in between the spines which complete the cage.

Venus's flytrap

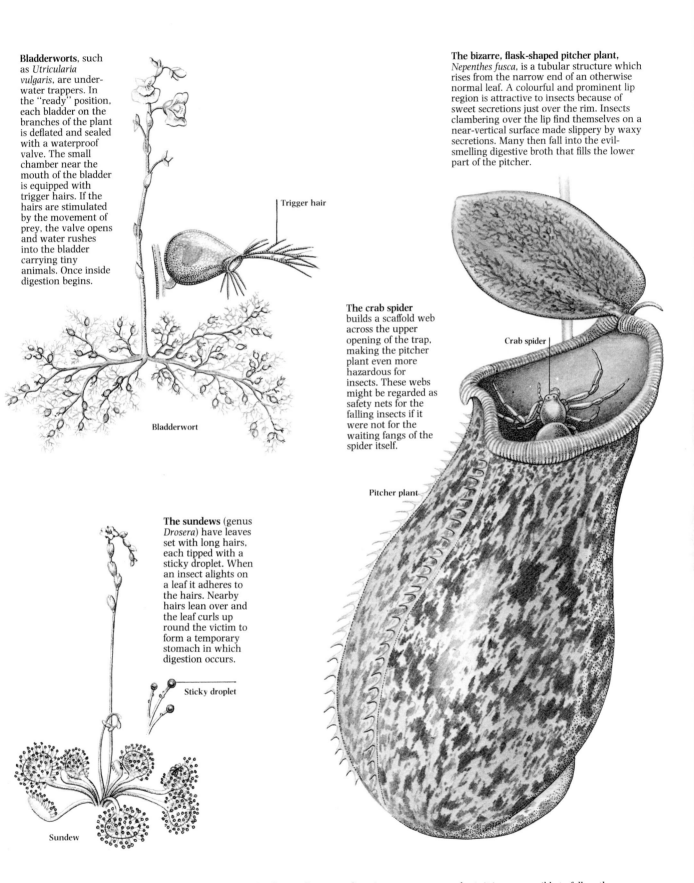

Bladderworts, such as *Utricularia vulgaris*, are under-water trappers. In the "ready" position, each bladder on the branches of the plant is deflated and sealed with a waterproof valve. The small chamber near the mouth of the bladder is equipped with trigger hairs. If the hairs are stimulated by the movement of prey, the valve opens and water rushes into the bladder carrying tiny animals. Once inside digestion begins.

Trigger hair

Bladderwort

The crab spider builds a scaffold web across the upper opening of the trap, making the pitcher plant even more hazardous for insects. These webs might be regarded as safety nets for the falling insects if it were not for the waiting fangs of the spider itself.

The bizarre, flask-shaped pitcher plant, *Nepenthes fusca*, is a tubular structure which rises from the narrow end of an otherwise normal leaf. A colourful and prominent lip region is attractive to insects because of sweet secretions just over the rim. Insects clambering over the lip find themselves on a near-vertical surface made slippery by waxy secretions. Many then fall into the evil-smelling digestive broth that fills the lower part of the pitcher.

Crab spider

Pitcher plant

The sundews (genus *Drosera*) have leaves set with long hairs, each tipped with a sticky droplet. When an insect alights on a leaf it adheres to the hairs. Nearby hairs lean over and the leaf curls up round the victim to form a temporary stomach in which digestion occurs.

Sticky droplet

Sundew

Once it was thought that carnivorous plants merely trapped their prey and depended on the secretions of enzymes by symbiotic bacteria on the leaf surfaces to achieve digestion. Now this is known to be incorrect. Carnivorous plants all have a rich supply of glandular areas with specialized cells. These cells secrete digestive enzymes and are sometimes involved in the absorption of food. The digestive fluid in the pitcher plant, for example, contains at least 2 potent, protein-splitting enzymes. One is similar to pepsin, the enzyme found in vertebrate stomachs. In many carnivorous plants it is now possible to follow the digestive enzyme release in great detail. The gland cells on the leaves of the butterwort, *Pinguicula*, for instance, are dry until stimulated by prey, but then respond almost instaneously by pouring out their digestive enzymes.

PREY DEFENCES: FIGHTING BACK

The interaction between predators and their prey is always two-sided: while predators are perfecting ever more efficient ways of finding and killing prey, the prey are engaged in the equally urgent evolutionary progression toward ever more elaborate defence mechanisms.

The multifarious means by which prey animals set up anti-predator defences are complex amalgams of structure, physiology, behaviour and group interaction. The successful anti-predator strategies that have actually been evolved include camouflage, mimicry, flight from attack, aggressive displays and retaliation, physical armour and cooperative tactics between animals of the same or different species. These defence mechanisms have been split into three broad categories—primary defences, secondary defences and defensive groups or associations.

Primary defences are mechanisms that go to work before a potential predator starts in pursuit or attack of its quarry. Their role is to decrease the likelihood that prey animals will have to fight off actual onslaughts by predators. Into this category come defences such as mimicry and camouflage, which may mean that the hunter does not detect its prey, confuses it with a harmful creature or detects it, but fails to recognize it as edible or palatable.

Only when a prey animal is actually attacked are secondary defences put into operation. These secondary defences act to increase the chance of an assaulted prey surviving the predatory encounter and may be passive or active. The plates and spines that arm sea urchins, or the protective shells of tortoises, are examples of the passive type, while active secondary defences include running away, threatening behaviour and active retaliation by attacking the predator with teeth, horns, hooves or other weapons.

In their struggle for survival, prey animals make use of two quite distinct sorts of group defences. Within family groups, or in herds, shoals or communities, individuals of the same species may cooperate to mount a shared defence against predators. This first sort of group defence may involve a mother bird sitting on her nestlings. Animal herds may take up defensive postures and some insect colonies even have defensive individuals whose specific job is to protect other community members. In the second type of communal defence one animal uses another organism of a different species as a mutual guardian. Such symbiotic associations require the combination of two living things to produce a difficult opponent: if left to their own devices, at least one of them would be easy pickings for a predator.

Animal defence tactics are paralleled in the plant world by a variety of anti-herbivore strategies. Some plants are protected by symbiotic link-ups with animals that ward off herbivores. A few plants attempt camouflage—by pretending to be rocks, for example. Some have damaging physical defences such as spines and thorns, and even more have evolved sophisticated techniques of chemical warfare by producing a biochemical armoury of toxins, including intensely injurious alkaloids. The alkaloids in a few plants of larkspur can kill a cow.

Display and Mimicry

Showy and ostentatious appearance or behaviour are defensive tactics used by many animals, both vertebrates and invertebrates. Whether passive or active, these displays can be vital to the war of nerves between predators and their prey.

Prey animals that have dangerous or unpleasant characteristics—such as being extremely unpalatable—often advertise the fact by having a striking, easily recognized appearance that notifies the predator of danger before it attacks. If such aposematic signalling appearances are to be worthwhile from the prey's viewpoint, the predator must

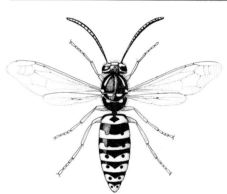

Wasp (yellow-jacket), *Vespula vulgaris*

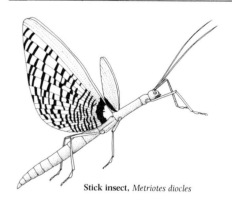

Stick insect, *Metriotes diocles*

Poisonous coral snake, *Micrurus fulvius*

Non-poisonous coral snake, *Lampropeltis doliata*

have sampled these noticeable prey, found them distasteful and remembered the experience. Alternatively, the avoidance of particular aposematic signals may be an instinctive, stereotyped part of predator behaviour built into its repertoire in the course of the animal's evolution.

Startling or deimatic behaviour is another sort of anti-predator display, but usually involves the sudden adoption of a posture, appearance or pattern of movement designed to intimidate the attacking predator. Such intimidatory displays may be truly threatening, in the sense that the prey animal is really able to defend itself to some purpose and harm the predator, but many prey species use this strategy as a bluff: they seem menacing but in fact lack the physical means of harming the predator in any way. The mimicker may however be just as successful.

Harmless prey may also fool their attackers by counterfeiting the looks or behaviour of a dangerous prey species with easily recognized aposematic signalling devices. For this Batesian mimicry to be successful both the copies and their harmful models must coexist so that predators can experience both sorts of animal.

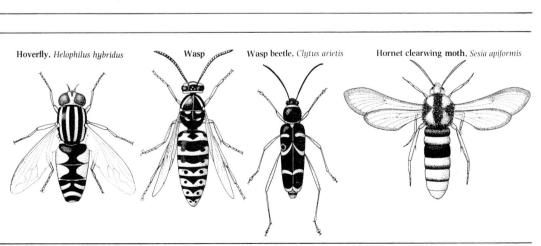

Skunk, *Spilogale putorius*

Ladybird, *Propylea punctata*

Electric ray, *Torpedo narke*

Advertising their noxiousness, the yellow and black striped wasps notify predators of their lethal stings. Striped or spotted in black and white, American skunks announce their ability to squirt out nauseous fluid. Red ladybird beetles, spotted in black, signal unpalatability, while the yellow-rimmed, blue-black spots of the electric ray warn of its electric shocks.

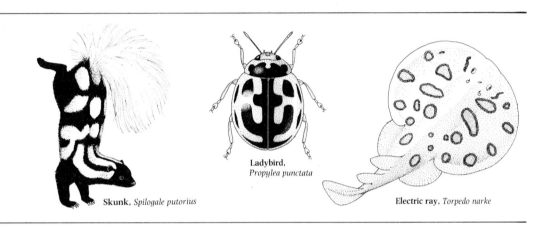

Carpet viper, *Echis carinata*

Toad, *Physalaemus nattereri*

Eye spot

Eye spot

Elephant hawkmoth caterpillar, *Deilephila elpenor*

Effective startle displays must come as a surprise. When provoked, a stick insect will rear up on its hind legs, fan its wings and face its attacker. The carpet viper makes a threatening display imitated by harmless snakes. Suddenly revealed, the eye-spots of the rump of the toad *Physalaemus* startle predators, as do the rapidly expanding spots on elephant hawk moth caterpillars.

Hoverfly, *Helophilus hybridus*   Wasp   Wasp beetle, *Clytus arietis*   Hornet clearwing moth, *Sesia apiformis*

By mimicking harmful animals, harmless prey gain a kind of camouflage. Such Batesian mimics are not difficult to see but are hard to identify. Deadly coral snakes, striped in bands of red, black and yellow, are mimicked by harmless colubrids. Similarly the warning yellow and black stripes of wasps and bees have been used as models by stingless hover-flies and beetles.

Physical Protection

To help escape the jaws of their predators, prey animals are endowed with a bewildering array of physical structures, habits and patterns of behaviour. This wide-ranging arsenal can be divided into passive defences and those that involve some action by the prey.

Structures that allow some passive resistance to predation operate either before the predator makes its onslaught—primary—or as secondary defences which only become assets to the prey after the predator has committed itself to an offensive. The most obvious examples of passive protective devices are the tough outer coverings of some prey animals. Many physically tough skeletal materials have been devised and employed to construct these armours. Mineralized shells are common and are usually hardened with calcium carbonate and calcium phosphate. Toughened or tanned structural proteins such as keratin and sclerotin, or the chitin that forms the outer covering of arthropods such as insects and spiders, are also resilient materials that potential predators find difficult to crack.

Prey animals—like the predators that seek them out—use camouflage or

Camouflage works well for two colour types of the European peppered moth. The blotched grey and white form is well hidden on lichen-covered trees, while the black form is almost invisible on trees sooted by industrial pollution. From 1850 to 1895 north-west Britain's proportion of black forms increased to 95 per cent of the total population, but grey-white forms still predominate in pollution-free areas today and are on the increase. The marine worm, *Urechis*, makes an extensive burrow system to feed in. This protective house is used by other animals like gobies, bivalves and crabs.

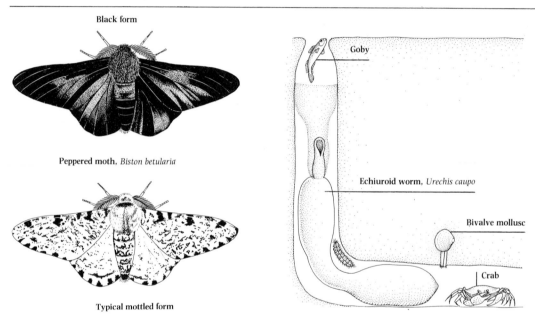

Black form

Peppered moth, *Biston betularia*

Typical mottled form

Goby

Echiuroid worm, *Urechis caupo*

Bivalve mollusc

Crab

Offensive weapons are essential if prey are to retaliate effectively. With horns and hard hooves, a large herbivore like an eland can often repel a pack of attacking hyenas. When assaulted by predators such as toads, the bombardier beetle puts a remarkable system of chemical warfare into action. The beetle's abdomen contains a sac filled with hydroquinone and hydrogen peroxide. The beetle squeezes this mixture into an "explosion chamber" where the two chemicals react to produce oxygen gas. This gas ejects a spray of the chemical mixture with a loud bang.

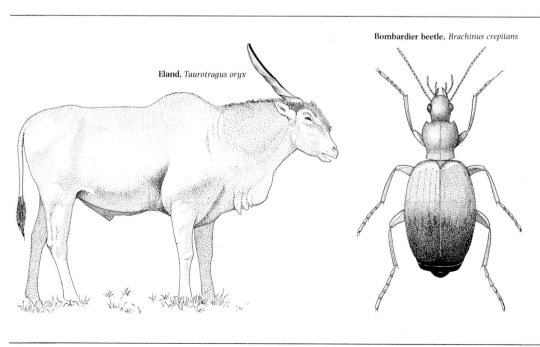

Bombardier beetle, *Brachinus crepitans*

Eland, *Taurotragus oryx*

crypsis techniques as passive defence measures. Such camouflage colouring makes a prey animal hard to see or to recognize.

Other forms of passive defence include fleeing rapidly from predators or hiding from them in tunnels, holes or cavities. Many marine creatures use the strategy of keeping out of sight, digging holes for themselves in the sea bottom, where they are safe from attack unless sought out by burrowing predators or probed from the sand by the long beaks of birds such as waders. The ultimate passive defence is feigning death, for this prevents the stereotyped killing response of the hunter from becoming activated. Several spiders and beetles use this trick, but its best known exponents are the American oppossums of the genus *Didelphis*—hence the expression "playing possum".

In contrast to these passive defences, many potential prey are quite capable of inflicting serious wounds on a predator brave enough to attack them. These active defences may be relatively simple ones, employing horns, claws or teeth, or more subtle methods of retaliation akin to chemical warfare.

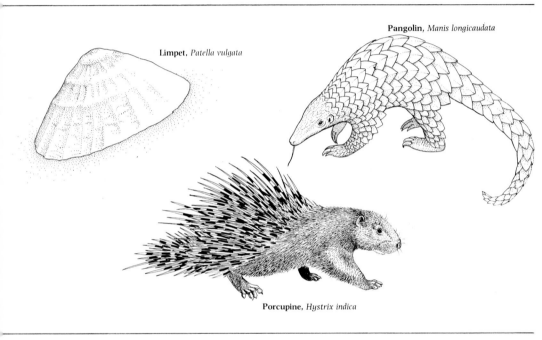

Limpet, *Patella vulgata*

Pangolin, *Manis longicaudata*

Porcupine, *Hystrix indica*

Tough outer cases that hamper predators are immensely varied in both invertebrate and vertebrate prey. Almost all molluscs have protective shells built from protein plus calcium salts. In limpets the shell is dual cover against wave pounding and attacks by fish at high tide. Stuck to a rock by its muscular foot, a limpet is impossible for man to remove with his bare hands. To deter potential predators many mammals have coats made from modified keratin, a skin protein. The spines of porcupines, and the fir-cone-like scales of pangolins are typical examples.

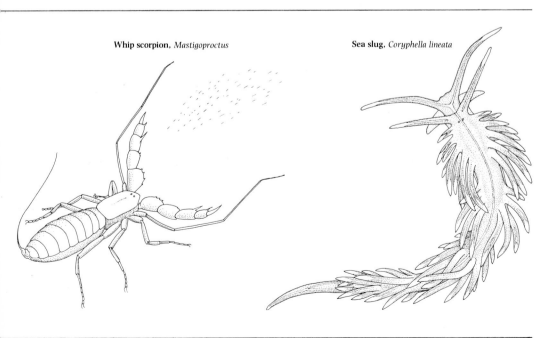

Whip scorpion, *Mastigoproctus*

Sea slug, *Coryphella lineata*

Sprays produced by glandular secretions and ejected at predators are the retaliatory equipment of arthropods such as cockroaches, bugs, stick insects and whip scorpions. If attacked, the whip scorpion *Mastigoproctus* rotates an opening of its anal gland to point at the aggressor and spurts out a fluid containing acetic acid. Some sea slugs are immune to the stinging cells of corals and sea anemones and can eat them whole. A few retrieve still-active stinging cells and transfer them to projections on their backs. When attacked, the sea slugs use their borrowed weapons.

Group Cooperation

In the continuous warfare between predators and prey, success depends not only on the structure and behaviour of the animals themselves but also on their environments and the other organisms that live in their vicinity. Cooperation between individual animals can help both the hunters and the hunted. For just as predators like wild dogs, lions and ants can kill a wider range of prey animals as groups than as individuals, so, in parallel fashion, prey animals have discovered the benefits of having communal defence mechanisms.

"United we stand, divided we fall" is the maxim that sums up the advantages of cooperation among prey animals against their attackers. Many prey species that habitually live in groups defend themselves and retaliate against predators by cooperative group efforts.

Even animals that do not normally congregate together will act in concord in the face of a common enemy. Song birds, for instance, will recognize each others' warning calls and, as a group, "mob" the adversary. Equally, all animals that have any family life—those that fend for or protect their young—will most probably act defen-

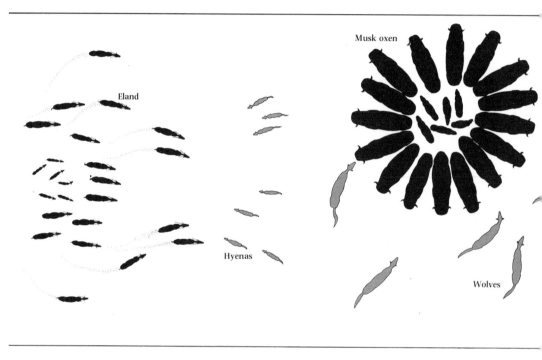

By acting in unison, eland of the Serengeti are rarely killed by spotted hyenas. As well as using horns and hooves to protect themselves against these predators, a herd of eland, when threatened by a hyena pack, takes up a distinctive group defence. The cow elands with young stay well back from the attackers while the cows without calves move forwards in retaliation. When menaced by a wolf pack, musk oxen form a defensive ring. With massive horns pointing outward and cows and calves in the ring's centre, the oxen can usually counter such an assault.

Eland

Hyenas

Musk oxen

Wolves

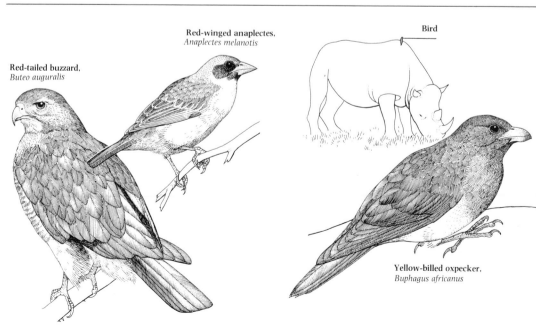

In choosing nesting sites many birds pick spots near animals whose defences they can use to advantage. In Africa small birds often nest in trees housing nests of large birds of prey. Because they do most of their hunting away from the nest, these carnivores are no danger to the smaller birds and many other predators steer well clear of the nest territory. Several birds associate with large African herbivores and act as their early warning systems. Tick birds, which feed on skin parasites of the herbivores, and egrets that live on insects disturbed by their feet, call when predators approach.

Red-tailed buzzard,
Buteo auguralis

Red-winged anaplectes,
Anaplectes melanotis

Bird

Yellow-billed oxpecker,
Buphagus africanus

sively as a family unit when their off-spring are threatened.

Communal defences are not, however, exclusive to animals of the same species, for the animal world displays a great variety of partly or wholly defensive relationships between members of different species. This mutually beneficial association of give and take is a form of symbiosis. The advantage accrued by each species in a particular symbiosis can be, for instance, the assurance of food supplies, but is often improved protection against predators.

The helping hand of symbiosis need not necessarily involve partnerships between animals—animals may provide plants with the vital extra in defence. In such a relationship a plant may, for example, gain protection from herbivores by associating with aggressive insects. A number of tropical plants attract ants into a close liaison by providing them with sugary nectar on stems, leaves and flowers or by developing cavities that can be used as nests. In both cases the resident ants, in return for this hospitality, use their jaws and stings to save the plants from the ravages of herbivores.

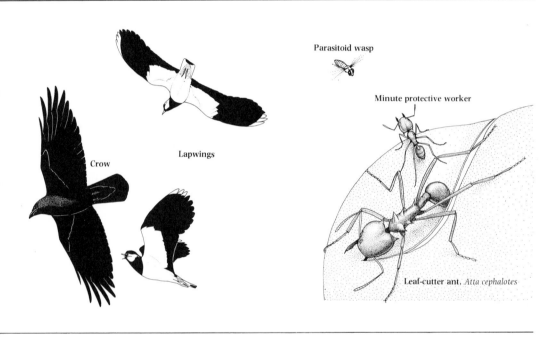

Crow

Lapwings

Parasitoid wasp

Minute protective worker

Leaf-cutter ant, *Atta cephalotes*

Flocks, herds, shoals or dense groups of a prey species can constitute a defensive advantage by distracting or confusing a predator. Lapwings will join forces to drive away crows which come near their nesting sites to prey on their young or their eggs. In colonies of some leaf cutter ants, large workers cut leaf segments and carry them back to the nest as food for the symbiotic fungi they rear. In this task the workers are susceptible to attacks by parasitoid wasps because their jaws are occupied, so they are accompanied by tiny minima workers whose task it is to ward off the attacking wasps.

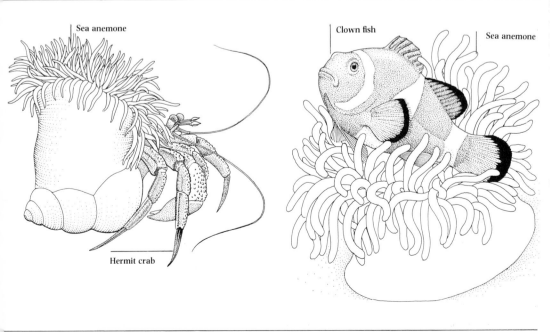

Sea anemone

Hermit crab

Clown fish

Sea anemone

Sea anemones, whose potent tentacles are armed with stinging cells, form intriguing associations with sea creatures, which gain protection by making use of these weapons. The sea anemones attached to many hermit crab shells probably save the crabs from attacks by large fish and octopuses. A crab may transplant the sea anemone on to its shell, or the anemone may somersault on to the crab's home. Clown fish often live among sea anemone tentacles. They are immune from harm because they coat themselves with mucus which inhibits the stinging cells' action.

Classification: how the animal world is structured

All the world's hunters, and the creatures on which they feed, have a place in the scheme of animal classification, which provides a key to the essential relationships within and be-. tween the groups of predators and prey whose lives and deaths are detailed in this book. The classification presented here is intended to make possible a reasonable working knowledge of all the groups of animals which are described in the preceding chapters.

Taxonomy — the science of classification — has arisen out of the need to attach accurate and concise names to the millions of different animals and plants that are recognized today. The common or vernacular names that form part of everyday vocabulary seem, at first glance, to meet these requirements, but if used worldwide such names can cause considerable confusion. The robin, for example, is a common European bird. Characteristically this small plump bird is about 5 inches (12 to 13 centimetres) in length and has long legs, olive-brown plumage on its back and a bright orange-red forehead, throat and breast. In North America, however, the bird that is called a robin is much larger — about the size of a blackbird, that is, some 10 inches (25 centimetres) long — has darker, almost black plumage on its back, a white ring round its eye and deep red feathering which is found only on its breast. In Australia the name robin is used for several kinds of flycatcher while in the West Indies it applies to the tiny, vivid green todies. All these "robins" have some red plumage on the throat or breast, but this feature alone is insufficient reason for calling them all by the same name.

To avoid the misunderstandings that can arise from using vernacular

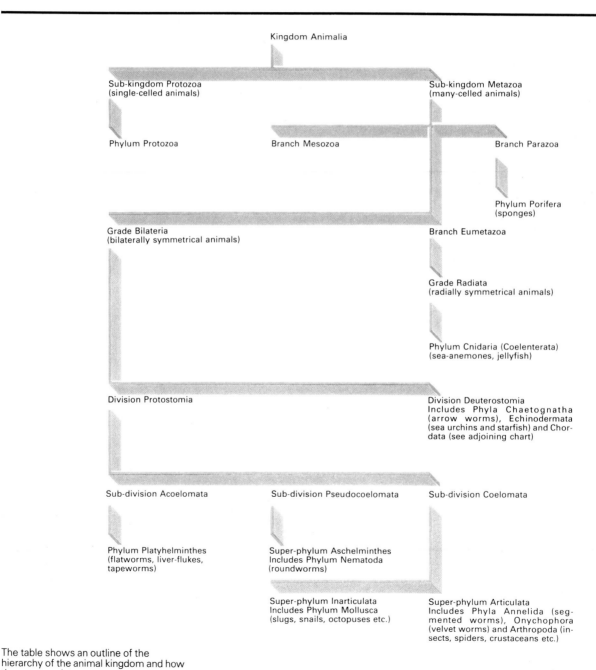

The table shows an outline of the hierarchy of the animal kingdom and how the groups relate.

names, taxonomists have provided animals and plants with alternative scientific names which also have the advantage of surmounting the barrier of language. The scientific name serves as an internationally accepted label for each kind or type of animal — that is, each animal species. The term species has proved rather awkward for scientists to define, but the most generally accepted modern view of the species is that it represents the members of a population of like kinds of animals which will reproduce and give rise to fertile offspring in their natural environment. The species can thus be thought of as the essential unit of animal (and plant) mating. The anatomical differences between members of the same species may sometimes be marked enough to justify their division into sub-species. These differences are not barriers to mating and often reflect geographical variations in coat or plumage.

Through detailed observation it is almost always possible to recognize that several species share common features and therefore form a natural higher grouping of related species. This grouping is called a genus. The two-word scientific names or binomials given to animals reflect these two fundamental grades of classification: the species itself and the group of similar species or genus. Taking the lion as an example, its scientific name is *Panthera leo*. The first word, *Panthera*, is the generic name and this is always written with a capital initial letter; the second, *leo*, is the specific name and is always written with a small initial letter. Both parts of the binomial are commonly printed in italics and no two genera may have the same name.

The genus *Panthera* describes a group of large, predatory cats which can be split into several species by virtue of their size, colouring, habits and geographical location. The genus includes not only the African lion but also the leopard, *Panthera pardus*; the Asian tiger, *Panthera tigris*; and the South American jaguar, *Panthera onca*. The specific name *leo* serves to distinguish the lion from all the other large cats and refers to the features that only lions possess such as a tawny yellow coat and the mane of thick hair round the male's head.

The patterns of relationships among animals were recognized in the seventeenth century and at that

Taking the lion as an example, the table, right, follows this species through all levels of classification.

Levels of classification		Associated characteristics
Kingdom	Animalia	All the animal forms
Sub-kingdom	Metazoa	Many-celled animals
Division	Deuterostomia	Term referring to position of mouth in the embryo of animals and the way the embryo develops.
Phylum	Chordata	Animals with a notochord (forerunner of the backbone) above which are a hollow nerve cord, post-anal tail and gill slits during some stage of development.
Sub-phylum	Vertebrata	Animals with backbones and a braincase.
Super-class	Gnathostomata	Animals with true jaws derived from gill-arch bones.
Class	Mammalia	Warm-blooded animals, suckle young on milk produced by mammary glands; only one bone (dentary) in lower jaw; three bones in middle ear.
Sub-class	Theria	Mammals whose young develop for some time in the female reproductive tract. Includes all mammals except the egg-laying monotremes.
Infra-class	Eutheria	Mammals whose young gain nourishment prior to birth from a placenta.
Order	Carnivora	Mostly flesh-eating animals (cats, dogs, bears, badgers, raccoons, weasels, etc.).
Sub-order	Aeluriodea	Cat-like animals including civets, cats, mongooses and hyenas.
Family	Felidae	Exclusively meat-eating cats; five digits on front paws, four on hind paws; all except cheetah possess retractile claws.
Genus	Panthera	Groups of closely related cats (lion, leopard etc.) distinguished by behaviour, habitat, coloration.
Species	Panthera leo	Lion. Tawny coloured coat. Hunts for large game animals in African savanna. Male has ruff on neck.

time, and in the century that followed, the study of these patterns was considered to be of paramount importance because they were presumed to reflect the principles and attitudes of the Christian God or Creator. Of all the classifications drawn up in this period, which did indeed seem to demonstrate some divine organizing force, by far the most important and useful was that of the Swedish naturalist Carolus Linnaeus. The father of modern taxonomy, Linnaeus invented the binomial system of labelling and in his *Systema Naturae,* whose tenth and most extensive edition was published in 1758, he formulated a scheme of classification that was to be the foundation stone for the one used today.

Despite the undoubted usefulness of such early classifications, the real reason why classifications seemed to demonstrate true relationships between animals and plants did not become apparent until the theory of evolution had been accepted. Although the idea that living things evolved, rather than being divinely created, was an extremely ancient one dating back to the Greeks, it was not generally believed to be true until after Charles Darwin published *The Origin of Species* in 1859. Once this theory was believed as an accurate reflection of the way in which animals and plants had come into being, it was obvious that the relationships shown up in classification were simply those of descent. The methods used by the earlier classifiers then came into their own, for in each case a large number of characteristics had been examined and species compared in the most minute detail. The differences could simply be incorporated into an evolutionary

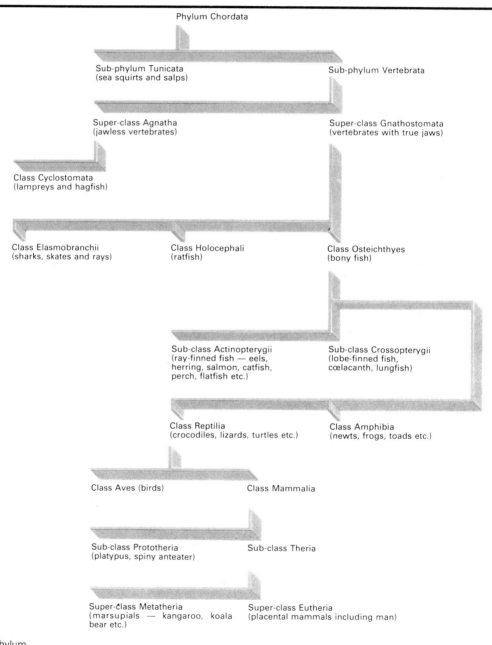

The table shows how the phylum Chordata is subdivided. It is composed largely of vertebrate animals.

classification as sequences in the descent of one species from another.

On the basis of evolutionary theory, and starting with the species, animals are classified today in a hierarchy of groups of increasing size. The number of features common to the animals and plants within each group decreases as these groups become larger and larger. So just as species are grouped together to comprise a genus, so genera with common characteristics and lines of descent are grouped to form a family and several families combined to compose an order. Related orders are combined as members of the same animal class and several classes make up the largest grouping usable for practical purposes, the phylum. All animal phyla are members of one huge group, the animal kingdom, which is made up of two large divisions, each containing many phyla. This method of classification thus forms a kind of taxonomic triangle, with the species occupying the apex and the entire animal kingdom stretched along its base.

Within the essential framework of this classification, scientists have found that some of the groups are too large to demarcate differences and relationships precisely. For this reason orders, classes, phyla and kingdoms are often subdivided into smaller groupings given prefixes to denote the nature of the division. Thus a sub-order is a group of related families representing some of the members of a complete order. Similarly a sub-class is a smaller group of orders than a class but larger than an infra-class. A group of several classes may, for example, be known as a super-class — an amalgamation smaller than a sub-phylum which is, in turn, smaller than a complete phylum.

For several reasons the higher levels of classification, that is, those above the level of genus, are not absolutely static. Changes are often made because increasing knowledge and the expansion of the fossil record may bring with them the realization that groups once thought to be closely related are in fact not related at all, or only distantly so. Another reason for change is that scientists differ in their interpretations of the best way for a classification to express relationships, but to prevent confusion, basic rules of taxonomy are laid down by the Commission of Zoological Nomenclature established in 1898. Genera and species may also sometimes have their names changed.

Phylum Protozoa
The simplest animals, consisting of a single, specialized cell. Many protozoans absorb food from the water in which they live, but some contain chlorophyll and produce their food by photosynthesis as plants do. Some are associated in colonies. The phylum therefore appears to occupy the borderline between the plant and animal kingdoms.

There are more than 30,000 species of protozoans, divided into four classes:

Class Mastigophora Free-living or parasitic protozoans characterized by the possession of at least one long, whip-like flagellum which beats to propel the animal through the water. *Chlamydomonas* is a common example.

Class Sarcodina Free-living or parasitic protozoans that engulf their prey by means of pseudopodia — extensions of the pliable body wall. A well known example of this class is *Amoeba*.

Class Sporozoa Parasitic protozoans. Among them is *Plasmodium vivax*, the organism that causes malaria. They usually have complex life cycles and are transferred from host to host by another animal called a vector e.g. the female *Anopheles* mosquito transfers the malarial parasite from the bloodstream of one human to another through its bite.

Class Ciliata Protozoans with bodies covered by many short, bristle-like hairs, called cilia. A very common ciliate is *Paramecium*, found in freshwater streams. It moves very rapidly by means of its cilia which beat in waves, and feeds on small particles which are drawn into the mouth down a special funnel lined with strongly beating cilia.

Phylum Porifera
The sponges. Animals with a body organization completely different from that of other many-celled animals. Sponges are separated off as a branch called the Parazoa. They consist essentially of two types of cell: some amoeba-like cells which build the tissue of the sponge and act as scavenger cells; the other cells are called collar cells and possess whip-like structures called flagella. The beating of the flagella draws a current of water and food particles in through pores in the sides of the sponge and out through an opening in the top. There are three classes of sponge:
Class Calcarea Sponges with skele-tons made entirely of chalk. The skeletal elements or spicules may be straight or much branched.

Class Hexactinellida The glass sponges. Their skeletons are composed entirely of silica; an example is the decorative Venus's flower basket (*Euplectella* sp).

Class Demospongiae Unlike members of the other sponge classes, these sponges have a jelly-like substance between the cells and a skeleton of a fibrous material (spongin) sometimes mixed with small amounts of silica. The bathroom sponge (*Euspongia mollissima*) is typical of this class. Its skeleton is particularly rich in spongin.

Phylum Cnidaria (Coelenterata)
Aquatic animals, the most primitive of the truly multi-cellular animals (**Eumetazoa**). They have radially symmetrical bodies (hence their grade: **Radiata**). The body has two cell layers, separated by a jelly-like substance, which enclose a body cavity with a mouth fringed by tentacles at one end. Surrounding the mouth is a whorl of tentacles bearing cells, cnidoblasts or nematocysts, which can seize, sting and paralyze prey. There are two basic body forms: polyps, which are cylindrical and remain attached to rocks (sessile), and medusae, free-swimming, umbrella-shaped forms. Some cnidarians pass through both medusoid and polyp phases in their life cycle. There are three classes:

Class Hydrozoa Most hydrozoans pass through both medusa and polyp stages in their life cycle. The mature form is the sessile polyp. Hydrozoans are inconspicuous creatures, but form much of the growth on rocks and shells. Many live in colonies of individuals, each housed in a thin tube.

Class Scyphozoa The jellyfish. The medusa is the predominant phase but there is a smaller, stationary polyp phase.

Class Anthozoa Solitary or colonial animals (sea anemones or corals) with no medusa phase. Corals are polyps protected by a hard skeleton of chalk.

Grade Bilateria
The bilaterally symmetrical animals (animals symmetrical about a vertical plane from head to tail) are organized in two large divisions, the **Protostomia** and **Deuterostomia**. These names refer to the way in which these animals develop from the embryo. The protos-

tomes are those in which the mouth develops from an aperture in the developing embryo called the blastopore. In the deuterostomes the mouth develops at the end of the embryo opposite the blastopore.

There are three sub-divisions of the protostomes: the **Acoelomata**, **Pseudocoelomata** and the **Coelomata**. Within the acoelomates there is one particularly important phylum:

Phylum Platyhelminthes
Flattened, worm-like animals. The body is unsegmented and has no true body cavity or coelom. Both male and female reproductive organs occur in one individual. Some are free-living, but many are parasitic. There are four classes:

Class Turbellaria Free-living flatworms that live in moist or wet environments and usually grow to 1¼ inches (3 centimetres) or less in length. *Planaria*, a common flatworm of fresh water, is a well known form remarkable for its powers to regenerate parts of its body.

Class Monogenea Parasites, many species of which live on the skin and gills of fish.

Class Cestoda The tapeworms. Parasites that live in the intestines of vertebrates. They have neither mouth nor intestine and absorb food directly through their body wall and remain attached to their hosts by special hooks or suckers.

Class Digenea Parasitic platyhelminthes such as the sheep liver-fluke (*Fasciola hepatica*).

Sub-division Pseudocoelomata Unsegmented worms having an alimentary tract with both a mouth and anus. There is a cavity around the gut but this is not a true coelom because it does not have a cellular lining. There are six classes, but only one of these will be described:

Class Nematoda The roundworms. All have long bodies, pointed at each end, covered by a thick skin or cuticle. There are parasitic and free-living species. *Ascaris* is a common roundworm.

Sub-division Coelomata Animals with true body cavities or coeloms, grouped into two super-phyla, the **Inarticulata** and the **Articulata** which are respectively unsegmented and segmented coelomates. The **Inarticulata** contains one particularly important phylum which is described:

Phylum Mollusca
Animals with bodies divided into a head, a muscular foot and a humped back covered by a mantle of skin which is folded to form a cavity; this mantle cavity acts either as a gill chamber, or as a lung in land-living forms. The mantle is able to secrete the animal's chalky shell. There are six classes of mollusc:

Class Monoplacophora The most primitive living molluscs. They resemble limpets and appear to show some segmented organization unlike the more advanced molluscs. The only genus is *Neopilina*.

Class Amphineura The chitons. Bilaterally symmetrical animals lacking eyes and tentacles. The well-developed foot adheres firmly to rocks on which they live around the intertidal zone of the sea shore.

Class Gastropoda The slugs, snails and limpets. The main part of the body is twisted. Most have coiled shells but in the slugs the shell has been gradually lost during the course of evolution. Gastropods move slowly by a creeping movement of the large foot.

Class Scaphopoda Tusk shells live in the sandy sea floor in long, pointed shells.

Class Lamellibranchia (Pelecypoda) The bivalves. Flattened molluscs with rounded or elongated shells and a mantle divided into two sheets that cover each side of the body. Each mantle secretes a shell and the two shells are hinged across the animal's back. The common mussel (*Mytilus edulis*) is a typical example.

Class Cephalopoda Squid, cuttlefish, octopuses. Extremely sophisticated animals with highly evolved eyes and nervous systems. The foot is modified into an array of tentacles with suckers. The shell is absent or greatly reduced. Cephalopods can swim by means of jet propulsion using a siphon to direct a jet of water forced out of the mantle cavity by muscular contraction.

Super-phylum Articulata includes five phyla, three of which will be described:
Phylum Annelida
Worms with bodies divided into repeating segments. The annelid body is sheathed in a thin, tough cuticle from which bristles (chaetae) protrude. At the front or head end a small brain forms a ring around the alimentary canal. The body cavity or coelom surrounds the gut and beneath this is the

nerve cord. There are three classes:
Class Polychaeta Marine worms with many bristles that grow from blunt, leg-like processes called parapodia. A pair of parapodia is found on each body segment, except round the head. Some are free-living, for example the predatory ragworm (*Nereis virens*); others are tube-living, filter-feeding worms such as the fan worm (*Sabella pavonina*).

Class Oligochaeta The earthworms. Animals that burrow in the soil and feed by swallowing soil and digesting the organic matter it contains. These worms have far fewer chaetae than the polychaetes and do not have parapodia. Each worm is hermaphrodite, possessing both male and female sexual organs, but to avoid self-fertilization they have a complicated mating procedure. The common earthworm (*Lumbricus terrestris*) is typical.

Class Hirudinea The leeches. Parasitic and predatory annelids. They have neither bristles nor parapodia, but have suckers at both ends of the body with which they cling to their prey. They are usually found in water. A typical example is the medicinal leech (*Hirudo medicinalis*) once used in medicine to draw blood.

Phylum Arthropoda
The largest phylum of the animal kingdom. The bodies of arthropods are segmented like those of annelids, but are encased in a horny cuticle which can be either flexible, stiff or rigid and forms an external skeleton (exoskeleton). Some of the body segments possess appendages, such as legs, wings and gills. There is always a head bearing sense organs and mouthparts. In the most primitive arthropods all the body segments possess simple appendages used as paddles for swimming. The evolution of the arthropods centres primarily around the specialization in the use of these appendages to perform separate functions as antennae, mouthparts, legs, gills, claws and so on. There are thirteen classes of arthropods, but only six have been covered in this book.

Class Crustacea Primarily aquatic arthropods which breathe through feathery gills; they have two pairs of antennae, and a body divided into head, thorax and abdomen although the head and thorax are usually indistinguishable. The head part bears three pairs of feeding appendages or mouthparts which are often aided by some of the anterior thoracic appendages. Most of their appendages are two-branched (biramous). Of the eight sub-classes, three are of particular interest:

Sub-class Copepoda Mostly tiny creatures no longer than $\frac{1}{4}$ inch (6 millimetres) that live in the plankton, feeding on minute diatoms. Some species are parasitic on freshwater and marine fish, living on the gills, fins or skin of fish and feeding on their blood and tissues.

Sub-class Cirripedia The marine barnacles. Young barnacle larvae are free-swimming and actively search for a suitable place, usually a rock, on which to settle. They attach themselves by an adhesive pad on their head and secrete a box of chalky plates for protection. Barnacles feed by protruding their long, feathery legs through the top of their plated box, and catching plankton in the water. A common European barnacle is *Balanus balanoides*. Parasitic barnacles can live on a variety of hosts including molluscs, corals, starfish and large crustaceans. *Sacculina*, a parasite of crabs, can be seen as a bright orange, spongy growth beneath the abdomen of crabs.

Sub-class Malacostraca The crabs, lobsters, prawns, shrimps and wood lice. Found in many habitats, from the sea (shrimps) to houses (wood lice).

Class Arachnida The spiders, scorpions, harvestmen, ticks and mites. Almost all are land animals. The body is divided into two regions, a prosoma (rather like head and thorax combined) and a hind body or opisthosoma. The prosoma bears six obvious paired appendages: four pairs of legs and in front of these a pair of chelicerae and pedipalps, which often bear large pincers. The opisthosoma bears no appendages. Arachnids are nearly all aggressive predators. There are ten orders, six of which will be described:

Order Scorpionida Secretive and nocturnal animals abundant in the tropics. The pedipalps of scorpions are developed into large pincers. The last segment of the body is modified as a sting which in some species is extremely venomous.

Order Pseudoscorpionida The pseudoscorpions. Small arachnids that live in leaf litter and damp corners of houses. They do not have stings, but have two pairs of pincers.

Order Solifugae Large swift-moving animals found in tropical regions. They have large, pincer-like chelicerae and long spindly pedipalps with sticky ends used for catching prey.

Order Araneae The spiders. Their chelicerae have poisonous fangs and their pedipalps are leg-like; in the male they are also modified as copulatory devices. The common garden spider, *Araneus diadematus*, is the most numerous species in Europe.

Order Opiliones The harvestmen or daddy-long-legs. Arachnids are characterized by compact bodies and very long legs. Their pedipalps and chelicerae are relatively small.

Order Acarina The mites and ticks. Many are parasites. In the parasitic forms the chelicerae and pedipalps are needle-like for piercing their hosts' skin.

Class Onychophora The velvet worms. Animals that live in rotting logs in the tropics. The body has a soft exoskeleton and short legs on each trunk segment. These extraordinary animals appear in many ways to be intermediate between annelids and arthropods.

Class Diplopoda The millipedes. Burrowing vegetarians. Each of the body segments is cylindrical and bears two pairs of legs.

Class Chilopoda The centipedes. Carnivorous, fast-moving arthropods. Unlike the millipedes, which they resemble, centipedes have a single pair of legs on each segment. The first pair of legs is modified into a pair of poison fangs.

Class Insecta Arthropods with a head bearing one pair of antennae and three pairs of mouthparts; a thorax of three segments, each of which bears a pair of legs, and usually a pair of wings on the two rear segments; and a legless abdomen. There are 29 orders, 13 of which are described:

Order Odonata The dragonflies and damselflies. Strong fliers with long, narrow abdomens, very short antennae, huge eyes and legs swung well forward to catch the small insects on which they feed. Their larvae are savage, aquatic predators.

Order Dictyoptera The cockroaches and praying mantids. Cockroaches are insects with long antennae, cigar-shaped bodies, and thick, leathery forewings which protect the hindwings. They are fast-moving, agile omnivores, often found in houses. The praying mantis and its relatives lie in wait for their prey which they grasp with their large, spiny front legs. The female frequently eats the male after mating.

Order Isoptera The termites. After a brief courtship flight, male and female termites mate and found a colony which can eventually number many hundred thousand. Their colonies are highly organized with workers to tend for the colony and soldiers to defend it from predators.

Order Orthoptera The grasshoppers and crickets. Insects with long hind legs used for efficient jumping. They create noises (stridulate) by rubbing their legs against the sides of their bodies. Locusts, which periodically swarm in large numbers destroying crops and vegetation, are typical orthopterans.

Order Dermaptera The earwigs. Insects with horny forceps on their tails which are used in defence. Short, leathery forewings protect the delicate hind wings which are folded when not in use. A common earwig is *Forficula auricularia*.

Order Phthiraptera Chewing and sucking lice. Chewing lice are tiny, wingless insects that live as parasites on birds and some mammals. They feed on feathers, skin and blood and are commonly called bird lice. The sucking lice are small parasitic lice of mammals with special mouthparts for piercing the skin to suck blood. The human louse (*Pediculus humanus*) is found worldwide.

Order Hemiptera An order including a variety of true bugs such as bedbugs, water boatmen, cicadas and aphids.

Order Neuroptera The lacewings and ant-lions, most of which are predatory insects. The adults have biting mouthparts and thin, membranous wings which are folded back over the abdomen when the insect is at rest. The larvae are predatory.

Order Coleoptera The beetles. Insects varying greatly in size and shape. Their forewings form horny sheaths completely concealing the hind wings.

Order Lepidoptera Butterflies and moths. Insects with two pairs of wings covered with powdery scales. The mouth has a long sucking proboscis which is coiled when not in use. The butterflies are daytime (diurnal) fliers that fold their wings vertically, whereas the moths are nighttime (nocturnal) fliers and fold their wings horizontally.

Order Diptera The true flies. These insects have only one pair of wings, the hind pair being reduced to small stumps which act as stabilizers. The group includes midges, craneflies, mosquitoes, horseflies, bluebottles and fruit flies. The housefly (*Musca domestica*) is a familiar example.

Order Siphonaptera The fleas. Parasites of warm-blooded animals —

mammals and birds. They are wingless insects with bodies flattened from side to side and large hind legs used for jumping on to their prey. The mouthparts are specialized for piercing and sucking blood.

Order Hymenoptera The ants, bees and wasps. Characterized by two pairs of glossy wings which are linked together. Some members of the order live in complex social groups similar to those found among termites. Ichneumon flies are members of this order. *Apanteles glomeratus* lays its eggs in the caterpillars of the cabbage white butterfly.

Division Deuterostomia (see p. 151)

Phylum Chaetognatha
The arrow worms. A small group of predatory, worm-like creatures that live in the plankton. The elongate body is divided into a head with eyes and two crescent-shaped jaws of horny teeth, a trunk and a tail, along the sides of which are fins.

Phylum Echinodermata
The echinoderms are the only higher invertebrates to exhibit fivefold radial symmetry. Echinoderms move slowly and have no head or brain to coordinate their activities. Their bodies are covered by chalky plates and spines embedded in the skin. There are five classes which are exclusively marine:

Class Asteroidea The starfish. Echinoderms that usually have five thick arms. They move by using small tube feet bearing suckers which are found in grooves underneath each arm. Starfish are usually carnivorous, feeding on bivalve molluscs. The notorious crown of thorns starfish (*Ancanthaster planci*) feeds on the polyps of reef corals.

Class Ophiuroidea The brittle stars. Echinoderms with slender arms used to pull the animal along the sea floor as the tube feet have no suckers. Brittle stars feed on plankton.

Class Echinoidea The sea urchins. Animals usually spherical in shape and without arms. The chalky plates form a rigid shell which is covered in a variety of spines. They move using tube feet and browse on algae growing on rocks.

Class Holothuroidea The sea cucumbers. Sausage-shaped animals with no arms. Along one side they have rows of tube feet with which they creep slowly across the sea floor. They feed on small particles of food on the sea bed which are caught with tentacles around the mouth.

Class Crinoidea The sea lilies. An ancient group of animals that live in deep seas and are permanently attached to the sea floor by a stalk. Some shallow-water forms become detached from the stalk when adult and swim by waving their feathery arms. They feed on small food particles caught on the arms.

Phylum Chordata
A large and highly successful group of deuterostomes, they are characterized by having a supporting skeletal rod or notochord, above which are a hollow nerve cord, gill slits and a post-anal tail. This group includes not only the vertebrates, but two other sub-phyla, the **Tunicata**, which will be briefly described, and the **Cephalochordata** whose only representative is the amphioxus or lancelet (*Branchiostoma lanceolatum*).

Sub-phylum Tunicata The tunicates consist of three animal classes, the best known of which are the sedentary Ascidiacea (sea squirts) and the Thaliacea (salps). Tunicates are unsegmented marine chordates that live inside a tough covering or tunic. The adults bear little resemblance to other chordates, but the larvae are tadpole-like and possess a notochord, dorsal nerve cord, gill slits and a post-anal tail. The adults feed by filtering sea water.

Sub-phylum Vertebrata The vertebrates are distinguished from other chordates by having a notochord strengthened or replaced by cartilage or bone, to produce a true backbone. The brains of vertebrates are protected by a skeletal cranium.
There are two super-classes of vertebrate, the Agnatha (jawless vertebrates) and the Gnathostomata (vertebrates with true jaws).
Super-class Agnatha (Cyclostomata) These were the first vertebrates. The few living descendants of the group belong to the order **Cyclostomata** including the lampreys and hagfish, which are parasites and scavengers. Hagfish are marine, eel-shaped fish which live on dead or dying fish, finding their prey by smell and touch. Lampreys have large sucker-shaped mouths by which they attach themselves to the flanks of fish. The tongue is covered in sharp teeth which cut into the flesh of the fish so that the lamprey can feed on its blood.

Super-class Gnathostomata The jawed vertebrates include all the remaining major groups of animals separated into seven classes: three classes of fish, **Elasmobranchii** or **Chondrichthyes**, **Holocephali** and **Osteichthyes** plus the **Amphibia**, **Reptilia**, **Aves** and **Mammalia**.

Class Elasmobranchii (Chondrichthyes) The cartilaginous fish, including the sharks, skates and rays. They have paired triangular fins which act as hydrofoils, provide lift and act as stabilizers. The mouths are adapted for biting, and whorls of continually growing teeth are attached to each jaw. The rough skin is covered by small, tooth-like scales. The sense of smell in these fish is very keen. The male fertilizes the eggs while they are still inside the female, using a pair of claspers that lie on the inner side of the pelvic fins. The sharks belong to the order **Selachii** and are excellent swimmers, with streamlined bodies. Most genera are predators. The rays and skates belong to the order **Batoidea** and are characteristically flattened, with wing-like pectoral fins.

Class Holocephali The rat-fish or chimaeras. An aberrant group of deep-sea fish that show some affinities with the elasmobranchs.

Class Osteichthyes The bony fish. The dominant fish of the seas and fresh waters. Fish with skeletons composed largely of bone. They have one gill opening on each side of the body, covered with a flap of tissue called the operculum, and a lung which is usually converted into a buoyancy tank or swim bladder.
There are two sub-classes:

Sub-class Crossopterygii The lobe-finned fish including the lungfish and coelacanth (*Latimeria chalumnae*). One order of this group, the **Rhipidistia**, gave rise to the first land-living vertebrates, the Amphibia.

Sub-class Actinopterygii The ray-finned fish. Fish with paired fins supported by jointed, bony rays. They have large eyes and no internal nostrils. All have swim bladders.
The sub-class **Actinopterygii** is divided into three infra-classes:

Infra-class Chondrostei Fish with thick scales with an outer layer of a shiny, enamel-like substance called ganoine. The mouth is elongated and the lower jaw hinged far back. The group includes the bichir of Africa (*Polypterus*) and the sturgeon (*Acipenser*) of the North Atlantic.

Infra-class Holostei Fish with ganoid scales like those of Chondrostei but shorter jaws. The two orders include the gar pikes (*Lepisosteus*) and bow fins (*Amia calva*) of North America.

Infra-class Teleostei The largest group of bony fish and prevalent in all waters. They evolved from the holosteans and have thin scales, without ganoine. The upper jaw is only very loosely attached to the skull so that the lips can be protruded. There are three basic divisions of modern teleosts. Selected examples are given of each division:

Division I includes the order **Anguilliformes**, the marine eels, including the conger eel.

Division II includes the order **Mormyriformes**, fish with toothed tongues found in rivers in Africa. They have tubular down-curved snouts.

Division III, by far the largest grouping, contains five super-orders and numerous orders, some of which have been selected:

Order Clupeiformes includes all the herrings which are migratory, shoaling fish.

Order Salmoniformes, the pike and salmon group of fish, all have a small fleshy fin behind the dorsal fin.

Order Myctophiformes are deep-sea lantern fish and have light-producing organs on the sides of their bodies.

Order Cypriniformes contains the majority of freshwater fish; all members of this order have a chain of small bones connecting the swim bladder and inner ear and have very acute hearing. Includes such diverse forms as the electric eel (*Electrophorus electricus*) and the piranha, (*Serrasalmus nattereri*).

Order Siluriformes are the catfish, most of which have feelers (barbels) around their mouths.

Order Batrachoidiformes or toadfish are predatory marine fish with large heads, strong teeth and long, tapering bodies.

Order Gobiesociformes or clingfish are able to cling to rocks using a sucker formed from the pelvic fins.

Order Lophiiformes, the angler or frog fish, have squat bodies and huge mouths. The first bony ray of the dorsal fin has a flap of flesh which is waved to lure prey towards the mouth of these fish.

Order Gadiformes includes the cod and its relatives.

Order Beryciformes, the squirrel and whale fish, have a series of prominent spines in front of dorsal and anal fins.

Order Atheriniformes, the flying fish, are small, rather slender fish with spineless fins.

Order Gasterosteiformes, the sticklebacks, sea horses and pipe fish, have long bodies encased in a bony armour and have narrow mouths.

Order Scorpaeniformes, scorpion fish, gurnards and bull heads, have rather large, ridged heads.

Order Perciformes includes the perches and their allies (mackerel and remora etc.). This is one of the largest orders of ray-finned fish.

Order Pleuronectiformes includes the flatfish such as plaice, sole, flounder and halibut.

Order Tetraodontiformes, the trigger and puffer fish, have tiny mouths with powerful "beaks"; their bodies are often covered with spines and their fins are small.

Class Amphibia The most primitive land-living vertebrates. Amphibia have not perfected the art of life on land because they are still dependent on water for breeding and require a moist environment to prevent their skins from drying out. The moist, thin skin of amphibians is used for respiration. Most amphibians develop by metamorphosis through a series of aquatic larval stages. There are three living orders:

Order Apoda The caecilians. Little-known legless amphibians of the Southern hemisphere, which closely resemble earthworms in their general body form.

Order Urodela. The newts and salamanders. Amphibians with legs and long muscular tails. Some adults have no lungs but respire through the skin or mouth alone. Larval characteristics are retained throughout life in some types, such as the mud puppy (*Necturus*).

Order Anura. The frogs and toads. Unlike the urodeles, anurans have loose-fitting skin, no tails when adult and the anatomy is specialized for jumping. Frogs and toads breathe through their skins and by raising and lowering the floor of the mouth, pumping air in and out of the lungs. *Rana temporaria* is the scientific name of the common frog.

Class Reptilia The first truly terrestrial vertebrates. Unlike amphibians, reptiles have thick, scaly skins. Their young do not pass through an aquatic larval stage but are born as miniature adults. They always breathe using lungs, never through the skin or with gills. Most reptiles lay shelled eggs in quantities from one to 1000, although some bear live young.

Reptiles cannot regulate their body temperatures like birds and mammals, but can maintain a fairly constant temperature by moving between warm and cool surroundings. There are four orders.

Order Chelonia (Testudines) The aquatic turtles and terrestrial tortoises. These reptiles possess a shell of bony plates covered with horny scales. The head and legs can be drawn inside the shell for protection. An extremely ancient group which has changed little in the 200 million years since it first evolved.

Order Rhynchocephalia The only surviving member of this order, the tuatara (*Sphenodon punctatum*), is now found only on islands off the north coast of New Zealand. Tuataras resemble lizards superficially, but their anatomy reveals that they are only distant relatives of the lizards.

Order Squamata The lizards and snakes. The most successful modern reptiles. Their bodies are covered in overlapping scales. There may be a single row of broad scales along the underside, as in snakes. The tongue is forked and the teeth are fused to the edge of the jaws. Some lizards are legless and resemble snakes. There are two sub-orders, the **Lacertilia** (lizards) and the **Serpentes** (snakes); they are not always easily distinguished, especially the snakes and legless lizards.

Sub-order Lacertilia Many lizards are snake-like in form with relatively small limbs. Some forms can shed the tail and regenerate it as a means of escaping from predators. Skin is shed at regular intervals and many lizards can change colour. The skinless lizards give birth to live young by retaining the fertilized egg inside the female. There are 20 families, of which four have been selected.

Family Chamaeleontidae The chameleons. Tree-dwelling lizards with feet specially adapted for grasping branches and prehensile tails. Their eyes can move independently. The tongue of the chameleon is extremely long and has a sticky pad at its end; it can be shot out to catch insects. The ability of chameleons to change colour for camouflage purposes is well known.

Family Lacertidae includes the common lizard (*Lacerta vivipara*).

Family Anguidae includes the slow worm, a legless lizard found in Europe (*Anguis fragilis*).

Family Varanidae The monitor lizards of the tropics. The largest is the Komodo dragon (*Varanus komodoensis*) which grows up to 12 feet (4 metres) long and is a predator and scavenger.

Sub-order Serpentes The snakes have mouths specially adapted for swallowing large prey, with loose and very flexible jaw joints. They are legless and their tails do not regenerate; only one of their lungs, the right one, is actually used — the left is reduced or absent. Snakes probably evolved from burrowing lizards at some period and are a very complex group to classify. Eleven families are recognized, two are described.

Family Boidae The constrictors. Large, non-poisonous snakes, which often have small remnants of hind limbs. They kill their prey by constriction, causing suffocation. Nearly all are found in the tropics. The largest species is the reticulated python (*Python reticulatus*) up to 40 feet (10 metres) long. The boa (*Constrictor constrictor*) lives in South America.

Family Viperidae The vipers. Snakes with tubular fangs which can fold back when the jaws are closed, but swing forward into a striking position when the mouth is opened. The fangs act like hypodermic syringes, injecting their venom into prey animals. The group includes: the common adder (*Vipera berus*) which feeds on small mammals; the rattlesnake (*Crotalus adamanteus*) of North America which grows a warning rattle on its tail tip from rings of shed skin.

Order Crocodilia The crocodiles and alligators. With the birds, the crocodilians are the closest living relatives of the dinosaurs. They are large predatory reptiles adapted for an aquatic existence. Features related to this life-style include a special valve at the back of the throat enabling animals to open their mouths under water and breathe with their nostrils above water; valves to close off ears and nostrils under water, and a powerful swimming tail. The body scales, called scutes, are large and horny, with plates of bone beneath. There are two families, the **Gavialidae**, long-nosed, fish-eating species from India and the **Crocodylidae**, including the broader-nosed crocodiles and alligators.

Class Aves (the birds) evolved from a group of reptiles seemingly closely related to dinosaurs. Bird feathers derive from reptilian scales, bird forelimbs are modified into wings and the thin, air-filled bones reduce their weight. Birds have no teeth, merely a horny beak. Like mammals, birds can maintain a constant body temperature. There are 28 bird orders, of which the **Passeriformes** contains more than half the known species.

Order Apterygiformes The kiwis. Small, flightless birds of New Zealand. They have short, stubby wings, no tail and short, strong legs. They feed at night, probing for insects with long, sensitive beaks.

Orders Struthioniformes, Rheiformes and **Casuariiformes** are respective orders of large, flightless birds: ostriches, rheas, and emus and cassowaries.

Order Podicipediformes The slavonian grebes. Weak fliers that spend most of their lives on water or feeding under water.

Order Gaviiformes The divers or loons. Arctic birds with streamlined bodies for diving.

Order Sphenisciformes The penguins. Southern hemisphere birds with wings which are converted into flippers with which they "fly" through the water.

Order Procellariiformes Albatrosses and petrels. Oceanic birds with long, tubular nostrils. They seldom come ashore except to breed, usually nesting in burrows in the ground.

Order Pelecaniformes Pelicans, gannets etc. The only birds with four webbed toes. Pelicans (*Pelecanus*) often feed in groups on shoals of small fish, using their beaks as scoops. This group of birds includes the frigate bird (*Fregata magnificens*) and the gannet (*Sula bassana*).

Order Ciconiiformes The herons, storks and flamingoes. Birds with stilt-like legs for wading, long bills, short tails and long, thin toes. They feed on fish and small crustaceans.

Order Anseriformes The ducks and screamers. All these water fowls have long necks and special oil glands at the base of the tail, which waterproof the feathers.

Order Falconiformes Birds of prey with sharp, hooked beaks. This group includes five families: the **Cathartidae** or vultures and condors of America which are scavengers rather than predators; the **Sagittariidae** or secretary birds which eat lizards and snakes; the **Accipitridae**, eagles, hawks and Old World vultures such as the Egyptian vulture (*Neophron percnopterus*), golden eagle (*Aquila chrysaetos*) and the Everglade kite (*Rostrhamus sociabilis*); the family **Pandionidae** includes the osprey; and the **Falconidae** includes the falcons, for example the peregrine falcon (*Falco peregrinus*).

Order Tinamiformes The tinamous. Terrestrial game birds of South America.

Order Galliformes includes most of the ground-living game-birds with heavy bodies such as grouse, turkeys and guinea fowl.

Order Gruiformes The cranes, rails and their allies. The cranes have long necks and legs and unwebbed feet.

Order Charadriiformes The waders, auks and gulls. Medium-sized shore birds, each possessing a tufted oil gland.

Order Pteroclidiformes The sandgrouses.

Order Columbiformes The pigeons and their allies.

Order Psittaciformes The parrots, parakeets, cockatoos and their relatives.

Order Cuculiformes The cuckoos. Reasonably close relatives of the parrots, but do not have hooked beaks.

Order Strigiformes The owls. Characteristically, owls have short, mobile necks, broad, flat faces with large eyes, and soft, fluffy plumage for silent flight. The barn owl (*Tyto alba*) has a wide distribution.

Order Caprimulgiformes The nightjars. Birds with long, pointed wings and large mouths. They feed on insects which are caught on the wing at dusk.

Order Coraciiformes An order that includes the kingfishers, hornbills and cuckoo-rollers. Carnivorous birds of worldwide distribution.

Order Piciformes The woodpeckers and toucans. Tree-dwelling solitary birds with two toes pointing forward and two pointing backward so as to grip vertical tree trunks. The green woodpecker (*Picus viridis*) is common in Europe and western Asia.

Order Apodiformes The swifts and hummingbirds. Short-legged birds capable of rapid flight.

Order Coliiformes The mousebirds. South African fruit-eating birds.

Order Trogoniformes The trogons. Insectivorous tropical birds.

Order Passeriformes The "perching birds". All have four unwebbed front toes and one hind toe. Their young are born naked and helpless. There is a very large number of families. The carrion crow (*Corvus corone*) is a typical member of this group which includes: larks, swallows, bowerbirds, birds of paradise, tits or chickadees, tree creepers, wrens, thrushes, shrikes, starlings, finches, sparrows and many others.

Class Mammalia Mammals are warm-blooded vertebrates. Their body temperature is regulated by a portion of the brain, and body hair and sweat glands help to maintain the constant temperature usually above that of the environment. Young are suckled on milk produced by the mother's mammary glands. Mammals evolved from a group of reptiles and one of the main anatomical changes involved the reduction of the lower jaw to a single bone (the dentary) and the conversion of two of the original jaw bones into middle ear bones. There are three principal types of mammal: the egg-laying monotremes; the marsupials, whose young are born very early and continue development in the mother's pouch; and the placentals, whose young remain within the mother's womb nourished by the placenta until after full gestation. Three-quarters of the living species of mammals are either rodents or bats.

Sub-class Prototheria The egg-laying mammals or monotremes of the order **Monotremata** are the most primitive living mammals. They have primitively organized brains, but manage to control their body temperature at a fairly constant level. They possess the beginnings of bones to support the pouch seen in the more advanced marsupials, although they possess no pouch themselves. Once the young have hatched they are nourished on milk, which they lap from small depressions on the chest of the mother. The male has a large horny spur on its ankle which is used for defence.

There are two families of monotreme: the first, the **Tachyglossidae**, includes the echidna or spiny anteater (*Zaglossus bruijni*) of New Guinea, Australia and Tasmania; the second, the **Ornithorhynchidae**, includes the one species of duck-billed platypus (*Ornithorhynchus anatinus*) of eastern Australia and Tasmania.

Sub-class Theria The mammals that do not lay eggs. There are two infra-classes.

Infra-class Metatheria The pouched or marsupial mammals including the order **Marsupialia**. The various families include:

Family Didelphidae, the opossums found in North and South America.

Family Peramelidae, the bandicoots found in Southeast Asia and Australia.

Family Phalangeridae, often called possums (not to be confused with the opossums) includes the koala bear (*Phascolarctus cinereus*).

Family Macropodidae, the kangaroos and wallabies.

Infra-class Eutheria The placental mammals. Classified in nineteen orders, thirteen are described:

Order Insectivora The insect-eating mammals. A group of primitive mammals with long, sensitive snouts, and sharp-pointed teeth for cracking open insect bodies. It includes animals such as the many species of shrew,

hedgehogs, moles, tenrecs and others.

Order Chiroptera The bats. The only mammals capable of true flight. Their wings are supported by long, thin, bony fingers; the wing membranes are extensions of the skin of the belly. There are two sub-orders of bat:

Sub-order Megachiroptera, the fruit-eating bats which are found in the tropics.

Sub-order Microchiroptera, are small bats with small eyes and short snouts; most are nocturnal and fly by means of echo location using very high-pitched cries. There are several families, including the **Rhinolophidae** or horseshoe bats and the **Desmodontidae** or vampire bats such as *Desmodus rotundus*.

Order Primates Mostly tree-dwelling mammals, showing a tendency toward grasping hands and feet bearing nails rather than sharp claws. Primate eyes face forward and they have good stereoscopic vision. Apart from man, who has a worldwide distribution, primates are tropical or sub-tropical. There are two sub-orders of primate. The **Prosimii** are the more primitive forms, with long snouts and eyes that do not face directly forward. This group includes the tree shrews, lemurs, lorises, bush babies and tarsiers. The other sub-order, **Anthropoidea** includes the higher apes, primates with short snouts and stereoscopic vision. These can be further divided into two groups: the New World monkeys of South America including the marmoset and spider monkey; and the Old World monkeys of Africa and Asia which include three families: the **Cercopithecidae**: colobus, langur, mandrill, rhesus monkey (*Macaca mulatta*) and the baboon (*Papio* sp); the **Pongidae** or apes (gibbon, orangutan, gorilla and chimpanzee); and the **Hominidae** or man (*Homo sapiens*).

Order Cetacea The whales, porpoises and dolphins. Mammals with streamlined bodies adapted for life in the water. They have no hind legs, the forelimbs are modified as flippers and the tail has a horizontal fluke. There are no external ears, although whales have very acute hearing. There are two sub-orders of whale: the **Odontoceti** or toothed whales, including sperm whales, narwhals, killer whales, porpoises and dolphins; and the **Mysticeti**, the baleen or whalebone whales including the blue whale and the Greenland right whale.

Order Edentata includes the anteaters, sloths and armadillos. A group of primitive New World mammals.

Order Pholidota The pangolins. Scaly anteaters of the Old World tropics.

Order Lagomorpha Pikas, rabbits and hares. Mammals that look rather like rodents, but have an extra pair of gnawing teeth. They are herbivores and have the peculiar habit of eating their faeces after each meal. This allows them to make best use of the nutrients in plant food.

Order Rodentia The gnawers. The most numerous of all mammals. They have continuously growing, chisel-like front teeth. Examples include squirrels, gophers, hamsters, rats and mice, gerbils, guinea pigs, porcupines.

Order Carnivora The flesh-eating mammals. Divided into two sub-orders, the **Aeluroidea** (cats, hyenas and civets) and the **Arctoidea** (dogs, weasels, bears, raccoons).

Sub-order Aeluroidea is divided into three families: the **Felidae**, including the lion (*Panthera leo*) the tiger (*Panthera tigris*) and the cheetah (*Acinonyx jubatus*); the **Hyaenidae**, laughing hyena (*Crocuta crocuta*); and the **Viverridae**, Egyptian mongoose (*Herpestes ichneumon*) and several types of civet.

Sub-order Arctoidea has four families: the **Canidae** or dogs, such as the jackal (*Canis mesomelas*), the cape hunting dog (*Lycaon pictus*) and the fox (*Vulpes vulpes*); the **Mustelidae**, the weasel family, including the stoat (*Mustela erminea*) the badger (*Meles meles*), otters, skunks and minks; the **Ursidae** or bears such as the brown bear (*Ursus arctos*); and the **Procyonidae**, the raccoons and pandas, for example the giant panda (*Ailuropoda melanoleuca*).

Order Pinnipedia includes the seals, sea lions and walrus. The walrus (*Odobaenus rosmarus*) lives in arctic waters. The upper canine teeth grow down to form large tusks.

Order Proboscidea The elephants of Asia and Africa.

Order Tubulidentata has one species, the termite-eating aardvark (*Orycteropus afer*).

Order Perissodactyla The odd-toed mammals. These hoofed mammals are all herbivores and, along with the artiodactyls, form many of the prey animals of the large predators. Includes the tapirs, rhinoceroses, horses, zebras and asses.

Order Artiodactyla The even-toed hoofed mammals. This order includes a large variety of herbivores such as pigs, peccaries, hippopotamuses, camels, deer, cattle and giraffes.

Index

An animal which is the subject of a main feature is indexed by use of two page numbers in bold type (eg Aardvark 72–3). Animals illustrated or discussed in subsidiary features have their page references in italics (eg Leopard 33). The classification (148–157) is self-indexing; but mention there of a creature discussed elsewhere in the book is in bold, single-page numbers (eg Ant 154).

Acknowledgements

The Publishers received invaluable help during the preparation of *The Hunters* from:
Ruth Binney, Marsha Lloyd and Sally Walters, who gave editorial assistance; Professor D. R. Arthur, who allowed zoological specimens from the Zoology Department of King's College, University of London, to be used as reference; Dr David Norman, who compiled the classification; Ann Kramer, who compiled the index.

The following books and articles were particularly valuable during preparation of *The Hunters: Anatomy and Mechanism of the Tongue of Chameleon carcaratus* C. P. Gnanamuthu, Proceedings of the Zoological Society 1930.
The Carnivores, R. F. Ewers, Weidenfeld and Nicolson, London 1973.

The Chordates, R. McNeill Alexander, Cambridge Univerity Press 1973.
Defence in Animals Malcolm Edmunds, Longman, London 1974.
The Feeding Habits of the Everglade Kite R. C. Murphy, Auk 72 1955.
The Honeyguides Herbert Friedman, Smithsonian Institute, Washington, 1955.
How Animals Run M. Hildebrand, Scientific American 1960.
In the Shadow of Man Jane van Lawick-Goodall, Collins, London 1970.
Innocent Killers Hugo and Jane van Lawick-Goodall, Collins, London 1970.
Invertebrate Zoology, P. A. Meglitsch, Oxford University Press 1972.
Kinetics of the Avian Skull W. J. Bock, Journal of Morphology 114 1964.

Looking at Animals Hugh B. Cott, Collins, London 1975.
A New Dictionary of Birds Sir A. Landsborough Thomson, Nelson, London 1964.
The Serengeti Lion George B. Schaller, University of Chicago Press 1972.
The Spotted Hyena Hans Kruuk, University of Chicago Press 1975.
Whales, E. J. Slijper, Hutchinson, London 1962.

Typesetting by Composing Operations Limited Tunbridge Wells, Kent.
Origination by Gilchrist Brothers Limited, Leeds.
Printed by Industria Gráfica Sa Barcelona, Spain.